Medical Intelligence Unit

Molecular Mechanisms of Ataxia Telangiectasia

Shamim I. Ahmad, BSc, MSc, PhD
School of Science and Technology
Nottingham Trent University
Nottingham, England

CRC Press
Taylor & Francis Group
Boca Raton London New York

CRC Press is an imprint of the
Taylor & Francis Group, an **informa** business

Molecular Mechanisms of Ataxia Telangiectasia

Medical Intelligence Unit

CRC Press
Taylor & Francis Group
6000 Broken Sound Parkway NW, Suite 300
Boca Raton, FL 33487-2742

First issued in hardback 2019

ISBN-13: 978-1-58706-322-0 (hbk)

Visit the Taylor & Francis Web site at
http://www.taylorandfrancis.com

and the CRC Press Web site at
http://www.crcpress.com

Library of Congress Cataloging-in-Publication Data

Molecular mechanisms of ataxia telangiectasia / [edited by] Shamim I. Ahmad.
p. ; cm. -- (Medical intelligence unit)
Includes bibliographical references and index.
ISBN 978-1-58706-322-0
1. Ataxia telangiectasia--Molecular aspects. I. Ahmad, Shamim I. II. Series: Medical intelligence unit (Unnumbered : 2003)
[DNLM: 1. Ataxia Telangiectasia--physiopathology. 2. Ataxia Telangiectasia--genetics. WL 390 M718 2009]
RC607.A83M65 2009
616.8'3--dc22

2009012990

Dedication

This book is dedicated to the sufferers of ataxia telangiectasia (AT) and their parents and relations who painstakingly look after their near and dears throughout their suffering periods. Also dedication goes to the AT Society of United Kingdom for their tireless work to try to eliminate the disease and to ease the suffering of the patients.

About the Editor...

SHAMIM AHMAD, after obtaining his Master's degree in Botany from Patna University, Bihar, India and his PhD in Molecular Genetics from Leicester University, England, joined Nottingham Polytechnic as Grade 1 lecturer and subsequently promoted to SL post. Nottingham Polytechnic subsequently became Nottingham Trent University where after serving for about 35 years he took early retirement to spend the remaining time in writing books and full time research. For more than three decades he worked on different areas of biology including thymineless death in bacteria, genetic control of nucleotide catabolism, development of anti-AIDS drugs, control of microbial infection of burns, phages of thermophilic bacteria and microbial flora of Chernobyl after nuclear accident. But his primary interest which started 25 years ago is the DNA damage and repair, particularly near UV photolysis of biological compounds, production of reactive oxygen species and their implications on human health including skin cancer and xeroderma pigmentosum. He is also investigating photolysis of non- biological compounds such as 8- methoxypsoralen+UVA, mitomycin C, and nitrogen mustard and their importance in psoriasis treatment and in Fanconi anemia. In 2003 he received a prestigious "Asian Jewel Award" in Britain for "Excellence in Education". He is also the Editor of *Molecular Mechanisms of Fanconi Anaemia* and *Molecular Mechanisms of Xeroderma Pigmentosum* published by Landes Bioscience.

CONTENTS

EDITOR

Shamim I. Ahmad
School of Science and Technology
Nottingham Trent University
Nottingham, England
Email: shamim.ahmad@ntu.ac.uk

CONTRIBUTORS

Note: Email addresses are provided for the corresponding authors of each chapter.

Melissa M. Adams
Department of Biochemistry and Molecular Biology
University of Texas Health Science Center
Houston, Texas, USA
Chapter 4

Lindsay G. Ball
Department of Microbiology and Immunology
University of Saskatchewan
Saskatoon, Saskatchewan, Canada
Chapter 10

Fred Bunz
The Department of Radiation Oncology and Molecular Radiation Sciences
Johns Hopkins University School of Medicine
Baltimore, Maryland, USA
Email: fbunz1@jhmi.edu
Chapter 7

Phillip B. Carpenter
Department of Biochemistry and Molecular Biology
University of Texas Health Science Center
Houston, Texas, USA
Email: phillip.b.carpenter@uth.tmc.edu
Chapter 4

Giuseppe Castello
Italian National Cancer Institute
CROM
Mercogliano (AV), Italy
Email: gbpagano@tin.it
Chapter 8

Paolo Degan
IST
Italian National Cancer Research Institute
Genoa, Italy
Chapter 8

Luitpold V. Distel
Department of Radiation Oncology
University Erlangen Nürnberg
Erlangen, Germany
Email: luitpold.distel@uk-erlangen.de
Chapter 6

David H. Dreyfus
Yale SOM
Founder Keren Pharmaceutical
New Haven, Connecticut, USA
Email: dhdreyfus@pol.net
Chapter 9

Tetsuo Nakajima
Research Center for Radiation Protection
National Institute of Radiological Sciences
Chiba, Japan
Email: otetsu@nirs.go.jp
Chapter 3

Susann Neubauer
Praenatalmedicine and Genetics Nürnberg
Nürnberg, Bavaria, Germany
Chapter 6

Giovanni Pagano
Italian National Cancer Institute
CROM
Mercogliano (AV)
Campania, Italy
Email: gbpagano@tin.it
Chapter 8

Ramune Reliene
Department of Pathology
UCLA School of Medicine
Los Angeles, California, USA
Chapter 11

Robert H. Schiestl
Department of Pathology
UCLA School of Medicine
Los Angeles, California, USA
Email: rschiestl@mednet.ucla.edu
Chapter 11

Steven M. Shell
Department of Biochemistry and Molecular Biology
James H. Quillen College of Medicine
East Tennessee State University
Johnson City, Tennessee, USA
Chapter 5

Akira Tachibana
Faculty of Science
Ibaraki University
Mito, Ibaraki, Japan
Email: akiratac@mx.ibaraki.ac.jp
Chapter 2

A.M.R. Taylor
CR-UK Institute for Cancer Studies
University of Birmingham
Birmingham, England
Email: a.m.r.taylor@bham.ac.uk
Chapter 1

Deborah Wilsker
The Department of Radiation Oncology and Molecular Radiation Sciences and the Sidney Kimmel Comprehensive Cancer Center
Johns Hopkins University School of Medicine
Baltimore, Maryland, USA
Email: dwilske1@jhmi.edu
Chapter 7

Wei Xiao
Department of Microbiology and Immunology
University of Saskatchewan
Saskatoon, Saskatchewan, Canada
Email: wei.xiao@usask.ca
Chapter 10

Yue Zou
Department of Biochemistry and Molecular Biology
James H. Quillen College of Medicine
East Tennessee State University
Johnson City, Tennessee, USA
Email: zouy@etsu.edu
Chapter 5

PREFACE

Ataxia telangiectasia (AT) is an incurable, rare autosomal recessive genetic disorder which affects approximately one in 40,000-100,000 and the carrier frequency is estimated to be 1:100-200. In the United Kingdom it affects 1 in 300,000 live births amounting to about 5 or 6 cases per year. Although the medial survival rate of affected persons is between 19-25 years, due to genetic heterogeneity, the survival rate can be fairly variable. In rare cases, however, patients survive into their third decade. The longest living patient on record died at age 34 years.[1]

Although the disease may have been identified earlier, according to PubMed the first published report on AT was in 1958.[2] These authors described the disease to be of a slowly progressive type; the patient suffered from sinopulmonary infection, which must have been due to a deficiency in the immune system, that was discovered later in AT patients. Since this publication, 4720 research papers including a number of excellent reviews have appeared (PubMed data, September, 2008). Results from these studies have provided exciting information on a multiplicity of hitherto unknown complex processes of sensing DNA damage, cell signalling, protein/protein interactions and their posttranslational processing, involving activities from control of redox state to repair of double strand DNA breaks (DSB), with outcomes that include apoptosis and carcinogenesis.

Since finalizing the chapters of this comprehensive treatise, a number of exciting findings have emerged and these have been integrated into this Preface.

Chapter 1 focuses on the clinical features of AT. The phenotype of AT usually appears by the age two years. In a recent study in Italy, although it is found that two AT siblings differed in clinical and immunological presentations,[3] clinically AT is characterised by multiple phenotypes including progressive neurodegeneration caused by cellular atrophy,[4] development problems of the upper and lower limbs, dysarthric speech, ocular telangiectasias ("spider veins") affecting movement disorders, awkward gait, slurred speech, and various other metabolic disorders. Ophthalmic features of AT include conjuctival telangiectasia, strabismus, saccadic dysfunction with headthrusts, and convergence insufficiency.[5]

In the last decade significant efforts have been put into understanding the molecular pathogenesis of AT in terms of enhanced susceptibility to DNA damaging agents, specifically those inducing double strand breaks (DSB) and their repair. These breaks are either repaired by non-homologous end-joining (NHEJ) or by homologous recombination (HR). Defect of any kind in the repair process can lead to chromosome instability and tumorigenesis. Severe DNA damage, beyond the capacity of the DNA repair processes, triggers apoptosis.[6] The cellular response to DSB is a highly coordinated and complex process involving a network of proteins that work as sensors, signal transducer mediators and effectors of repair.

Major molecular and physiological features of AT patients include immunodeficiency and proneness to cancer. Also cells from AT patients show increased radiosensitivity and chromosomal instability. The chromosomal instability is thought to be triggered either by misrepair of DNA or impaired DNA damage processing. ATM is suggested to maintain the stability by suppressing DSB-induced translocation events. Furthermore, the onset of cancer and normal tissue injury are mainly associated with chromosomal instability. Increased immunodeficiency usually leads to increased susceptibility of patients to infection. IgA deficiency and T and B cell lymphopaenia have also been detected; the latter may be due to low thymic output. A predisposition of breast cancer in women appears to be more common in AT patients.[7]

All the features described above have been associated with a deficiency of the cells to repair DSB, and this is due to a mutation in ATM gene. ATM is a large gene, localised to chromosome 11q22-23 by linkage analysis and by positional cloning.[8] The protein encoded by this gene, usually referred to as ATM protein kinase, belongs to the phosphoinositide-3- kinase (PIK) superfamily. The enzyme phosphorylates a number of proteins but not lipid. The ATM gene is 750 Kb encoding 3056 amino acids constituting a 370 kDa protein. It has a number of different domains and functions. Interestingly two other family members belonging to the PIK family are DNA- dependent protein kinase (DNA –PK) and Rad3 related (ATR) kinase; these also participate in DNA damage responses.

Interestingly, AT shares a number of clinical and molecular features with a variety of other syndromes, namely ataxia telangiectasia-like disorders (ATLD) caused by a mutation in a separate gene, hMRE11. Other AT-associated disorders include autosomal recessive cerebral ataxia (ARCA) that includes Freidreich ataxia, ataxia oculomotor apraxia Type 1 and Type 2 and Nijmegan breakage syndrome (NBS). NBS patients have mutation in the NBS gene and do not show cerebral ataxia.

Chapter 2 targets the genetics of the ATM gene including its mutations and heterogeneity in subgroups. It is interesting to note that mutations occurring in different regions of the ATM gene can manifest themselves in phenotypes from mild to severe, depending upon where in the gene mutation has occurred. Moreover, its effects on the ATM kinase range from truncation to heavy destabilization; other mutations result in reduced but detectable levels of the protein; the latter group of patients show milder forms of AT. Other features of AT mutations, including complementation analysis, AT variations and the molecular nature of mutations in different patients, have been intricately covered in this chapter.

From a wealth of studies it is clear that AT cells are defective in DNA repair, specifically of DSB, and show enhanced sensitivity to ionising radiation and radiomimetic drugs such as bleomycin.[9] It is, however, interesting to note that there are other known agents that can lead to DSB (such as mitomycin C, 8-methoxypsoralen and ultraviolet light A, cisplatin and nitrogen mustard), but fewer studies on these agents are available. In fact in one case it has been shown that, while cisplatin increases ATR activity, it decreases ATM and DNA-PK activities.[10] Perhaps extending studies with these other agents may shed further light on the molecular nature of DNA of the repair deficiencies in cells from AT patients. This kind of study may also lead to development of tools for simpler diagnosis among AT subgroups.

Chapter 3 is dedicated to describing the roles of ATM in controlling cell cycle arrest (G1-S, intra-S-Phase, and G2M that provides time for repair), once damage signals are relayed.[11] Soon after DSB are recognised, DNA replication and mitosis are halted to prevent the replication and segregation of damaged chromosomes. Subsequently a cascade of proteins, including Cdks, Cdks binding cyclin protein subunit, cyclin dependent kinase inhibitors (Cdkis) and Cd25 phosphatase A, B and C come into action. A large scale proteomic analysis of proteins phosphorylated in response to DNA damage has identified over 700 proteins recognised by ATM and ATR.[12] How some of these well- studied proteins participate in cell cycle arrest is highlighted in this chapter. The existence of such a large number of proteins indicates the complexity of the DNA processing pathways. Following this, the repair pathway comes into action. If successful repair cannot occur, the apoptotic pathway is taken in which ATM also plays a key role. Incorrectly repaired or unrepaired DNA can also promote tumorigenesis via this pathway.

ATM also plays roles in cell cycle signalling and apoptosis by involving protein kinase C, ceramide and JNK. In addition ATM regulates a variety of other cellular functions through the cell cycle such as replication origin firing, telomere dysfunction and programme events at antigen replicator loci in T and B cells of lymphoid origin. Chapter 3 addresses in depth the cell signalling issues in AT. In a recent study Godarzi et al have shown that heterochromatic DSB are repaired much more slowly and that euchromatic DSB and ATM signalling is specifically required for the former type of DSB. Evidence is presented that knockdown of the transcriptional repressor KAP-1, an ATM substrate, or the heterochromatin building factors, HP1 or HDAC1/2 alleviates the requirements for ATM for DSB repair.[13]

Adams and Carpenter in Chapter 4 have presented comprehensive information about how DSB is repaired in normal cells and affected in its repair in AT patients. As noted above, the product of ATM gene, the protein kinase, is responsible for initially recognising DSB. Once DSB is recognised, one of two pathways is used for its repair: non-homologous end joining (NHEJ) or recombinational repair. NHEJ is initiated by recognition of DSBs by the DNA end-binding heterodimer, Ku, and the final step of DNA end joining is accomplished by the XRCC-4 DNA ligase 4 complex. A recent in vitro study has shown that aprataxin and PNK-like factor (APLF), an endo/exonuclease with an FHA domain and unique zinc finger, interact with both Ku and XRCC-4 DNA ligase 4 in human cells. The interaction of APLF with XRCC-4 DNA ligase 4 is FHA- and phosphor-dependent, and is mediated by CK2 phosphorylation of XRCC4. The APLF phosphorylation occurs at serine residue 116.[14] Moreover, Oloffson et al have shown that activation of ATM induces Sp1 phosphorylation at serine 101.[15]

For DSB to be repaired, a complex pathway is triggered in which phosphorylation of a number of proteins occurs; these include histone H2AX (a variant of the H2A protein family that after phosphorylation becomes gamma-H2AX),[16] the mediator of damage checkpoint proteins, nibrin (the product of NBS1), p53 binding protein and breast cancer protein 1.[17] MRE11-NBS1(XRS2)-RAD50 complex (also known as MRN complex), a major sensor of DNA breaks, recruits ATM which is activated to phosphorylate members of that complex and a variety of other proteins involved in cell cycle control and DNA repair.[18] NBS1 has been shown to play a key role in identification and regulation of DNA damage repair and recent studies have identified interactions with more proteins, such as BAX and caspase 3.[19] These authors have shown that Ku70 also interacts with BAX. In another recent study it has been shown that MDC1 directly binds to NBS1 and targets the complex to DNA damage sites.[20] Aven (an apoptosis inhibitor) has been shown to act as promoter of autophosphorylation of ATM in response to DNA damage in human cells.[21] Yet another protein, WRN (Werner syndrome), activates ATM.[22] From these studies it is clear that the DNA damage response and repair are highly complex processes, and we are a long way away from complete understanding of the intricate mechanisms involved. MRN dependent processing of DSB leads to the accumulation of short single-stranded DNA oligonucleotides (ssDNA oligos); the MRN-bound to ssDNA oligos then stimulates ATM activity.[23] Other proteins involved in this process are: BRCA1, PARP-1, RAD18 and DNA-dependent protein kinase catalytic subunit (DNA-PKcs). Also phosphorylated are p53A, chk2 (a cell cycle effector protein) and NBS1 (the Nijmegan breakage syndrome) proteins. It also controls the transcriptional activity of NF-κB and p53, after exposure to DNA damaging agents. NF-κB also plays a critical role in response to biological genotoxic stimuli: its activation can be mediated by ATM-dependent phosphorylation of NEMO (NF-κB essential modulator) following DSB induction.[24] Further details of this complex process can be seen in Chapter 9.

Chapter 5 presents the intricate interactions of various proteins involved in ATM regulated actions. Of specific interest is its Table 1 where the names and actions of 24 different proteins are given. ATM and ATR have been shown to play key roles in complex interactive process.

Chromosomal mutation and instability is a major event observed in cells derived from AT patients. Chapter 6 provides a detailed analysis of this feature ranging from the types of mutation, the spontaneous frequency and the frequency induced by radiation in various phases of growth. A variety of single stranded DNA binding proteins (SSBs) have been discovered that play important roles in DNA replication, recombination, DNA damage detection and repair. In a recent study a new SSB protein has been identified (hSSB1) that has been shown to be phosphorylated by ATM kinase in response to DSB. This phosphorylation is required essentially for DNA-induced stabilisation of hSSB1. Upon stabilisation, the protein accumulates in the nucleus and forms distinct foci. Cells deficient in hSSB1 exhibit increased radiosensitivity defective checkpoint activation and enhanced genomic instability coupled with minimised capacity for DNA repair.[25]

Chapter 7 describes what happens when DNA sustains heavy damage and the repair systems fail to accurately process the damage; in this case the cell is directed towards apoptosis, a critical cellular function for cancer prevention.

In Chapter 8 Pagano and his colleagues have argued that oxidative stress also plays important roles in the development of the AT phenotype. Evidence has been presented which points to an endogenous pro-oxidant state in the AT phenotype. ATM protein kinase has been cited to be involved in regulating cellular redox homeostasis and in modulating the expression of proteins with antioxidant functions. Figure 2 in the chapter neatly highlights the central roles oxidative stress plays in determining the various cellular and clinical abnormalities observed in AT.

Chapter 9 is dedicated to describing oncogenesis in AT: the roles of ATM, NF-κB and DDE in recombination pathogenesis. It highlights ATM interactions, DNA hairpin structure and related DSB, generated through V(D)J recombination. ATM is known to stabilise chromosomal V(D)J recombination DSB intermediates.[26] It also facilitates chromosomal end joining and prevents broken DNA ends being converted to deletions, inversions and translocations. V(D)J recombination has also been proposed[27] to be involved in oncogenic recombination events of immunoglobin and T cell receptor genes, initiated by RAG-1 and RAG-2 and subsequently processed by DNA pk, Ku and Artemis, which can lead to oncogenesis via illegitimate recombination that may bring oncogenes into altered regulatory sites or produce abnormal oncogene directed fusion protein.

Chapter 10 describes the overlaps between AT, NBS and ATLD. Although these are closely related human diseases, nevertheless, often it becomes difficult when clinical diagnosis is sought. Intricate molecular analyses continue to be important tool for a clear cut diagnosis and hence this chapter prove important for clinicians working in this field.

Genetically constituted animals, with relevant genetic defects, have played important roles in understanding various genetics, biochemical and physiological defects in humans. To study AT, a variety of AT mutant mice have been produced and studied. In Chapter 11 Reliene and Schiestle have elaborately catalogued the large number of available mutant mice and their molecular natures and phenotypes. Of interest are the mice constituted in their own laboratory and the pioneering work they are carrying out.

Despite a huge volume of research already carried out, we are still some distance away from a complete understanding of the molecular mechanisms of AT, and there is no evidence that interest in this disease is waning. Hence this book should provide both expert and novice researchers in the field with an excellent overview of the current status of research and pointers to future research goals.

Shamim I. Ahmad, BSc, MSc, PhD

References

1. Opeskin K, Waterston J, Nirenberg A et al. Ataxia telangiectasia with long survival. J Clin Neurosci 1998; 5:471-473.
2. Centrewall WR, Miller MM. Ataxia, telangiectasia, and sinopulmonary infection; a syndrome of slowly progressive deterioration in childhood. AMA J Dis Child 1958; 95:385-396.
3. Soresina A, Meini A, Lougaris V et al. Different clinical and immunological presentation of ataxia telangiectasia within the same familiy. Neuropediatrics 2008; 39:43-45.
4. Kulkarni A. Wilson DM 3rd. The involvement of DNA damage and –repair defects in neurological dysfunction. Am J Hum Genet 2008;82:359-366.
5. Khan AO, Oystreck DT, Koenig M et al. Ophthalmic features of ataxia telangiectasia-like disorder. J AAPOS 2008; 186-189. Epub 2007 Dec 21 Links.
6. Morio T, Kim H. Ku, Artemis, and ataxia telangiectasia-mutated: signalling networks in DNA damage. Int J Biochem Cell Biol 2008; 40:598-603. Epub 2007 Dec 24 Links.
7. Mavrou A, Tsangaris GT, Roma E et al. The ATM gene and ataxia telangiectasia. Anticancer Res 2008; 28:401-405
8. Shiloh Y, Ataxia telangiectasia and the Nijmegan breakage syndrome: related disorders but genes apart. Annu Rev Genet 1997; 31:635-662
9. Cohen MM, Simpson SJ, Pazos L. Specificity of bleomycin-induced cytotoxic effects on ataxia telangiectasia lymphoid cell lines. Cancer Res 1981; 41:1817-1823.
10. Yazlovtskaya EM, Persons DL. Inhibition of cis-platin induced ATR activity and enhanced sensitivity to cisplatin. Anticancer Res 2003; 23:2275-2279.
11. Cann KL, Hicks GG. Regulation of the cellular DNA double-strand break response. Biochem Cell Biol 2007; 85:663-674.
12. Matsuoka S, Ballif BA, Smogorzewska et al. ATM and ATR substrate analysis reveals extensive protein networks responsive to DNA damage. Science 2007;316:1160-1166.
13. Goodarzi AA, Noon AT, Deckbar D et al. ATM signalling facilitates repair of DNA double-strand breaks associated with heterochromatin. Mol Cell 2008; 31:167-177.
14. Macrae CJ, McCulloch RD, Ylanko J et al. APLF (C2orf13) facilitates nonhomologous end-joining and undergoes ATM-dependent hyperphosphorylation following ionising radiation. DNA Repair (Amst) 2008; 7:292-302.
15. Olofsson BA, Kelly CM, Kim J et al. Phosphorylation of Sp1 in response to DNA damage by ataxia telangiectasia-mutated kinase. Mol Cancer Res 2007; 5:1319-1330.
16. Kuo LJ, Yang LX. Gamma-H2AX – a novel biomarker for DNA double-strand breaks. In Vivo 2008; 22:305-309.
17. Riches LC, Lynch AM, Gooderham NJ. Early events in the mammalian response to DNA double-strand breaks. Mutagenesis 2008. Epub ahead of print.
18. Lavin MF. ATM and MRE11 complex combine to recognize and signal DNA double-strand breaks. Oncogene 2007; 26:7749-7758.
19. Iijima K, Marunaka C, Kobayashi J et al. NBS1 regulates a novel apoptotic pathway through BAX activation. DNA Repair (Amst) 2008; July 30. Epub ahead of print.

20. Wu L, Luo K, Chen J. MDC1 regulates intra-S-phase checkpoint by targeting NBS1 to DNA double-strand breaks. Proc Natl Acad Sci USA 2008 Aug 4. Epub ahead of print.
21. Guo JY, Yamada A, Kajino T et al. Aven-dependent activation of ATM following DNA damage. Curr Biol 2008; 18:933-942.
22. Cheng WH, Muftic D, Muftuoglu M et al. WRN is required for ATM activation and the S-phase checkpoint in response to interstrand crosslink-induced DNA double strand breaks. Mol Biol Cell 2008; July 2. Epub ahead of print.
23. Jazayeri A, Balestrini A, Garner E et al. MRE11-Rad50-Nbs1-dependent processing of DNA breaks generates oligonucleotides that stimulate ATM activity. EMBO J 2008; 27:1953-1962. Epub 2008 Jul 3 Links.
24. Wu ZH. Miyamoto S. Induction of pro-apoptotic ATM-NF-kappaB pathway and its repression by ATR response in replication stress. EMBO J 2008; 27:1963-1967. Epub 2008 Jun 26.
25. Richard DJ, Bolderson E, Cubeddu L et al. Single-stranded DNA binding protein hSSB1 is critical for genomic stability. Nature 2008; 453:677-681. Epub 2008 April 30.
26. Bredemeye AL, Sharma GG, Huang CY et al. ATM stabilizes DNA double-strand-breaks complexes during V(D)J recombination. Nature 2006; 442:466-470.
27. Callen E, Jankovick M, Difilippantonio S et al. ATM prevents the persistence and propagation of chromosome breaks in lymphocytes. Cell 2007; 130:63-75.

Acknowledgements

The editor is thankful to Professor Fumio Hanaoka of Graduate School of Frontier of Biosciences, University of Osaka, Japan for providing a visiting Professorship at this university. It was during the stay there the idea came of producing this book and the preliminary preparation had started. The editor is also grateful to all the contributors for their excellent chapters.

CHAPTER 1

Clinical Features of Ataxia Telangiectasia

A.M.R. Taylor*

Introduction

Ataxia telangiectasia (AT) is a remarkable disorder, the basis of which is a deficiency in the cellular response to DNA strand breaks. It was the first human disorder to be described in which patients were shown to be unusually sensitive to ionizing radiation, a response synonymous with a deficiency in the repair of DNA double strand breaks. The reason for this was mutation of ATM, a large gene that encodes a 370 kDa protein kinase with important functions in the cellular response to DNA damage. The relationship of all the clinical features of AT to this basic cellular defect are still incompletely resolved and this is particularly true of the cardinal clinical feature of AT, the progressive neurodegeneration caused by cerebellar atrophy and possibly other brain lesions. Alongside AT must be considered ataxia telangiectasia like disorder (ATLD), where again patients show an unusual sensitivity to ionizing radiation and indeed as the name suggests, share many of the neurological and cellular features of AT. However, this disorder is caused by mutation of another gene, hMRE11. There are several other autosomal recessive cerebellar ataxias (ARCAs) that also include Friedreich's ataxia, ataxia oculomotor apraxia Type 1[1-3] and ataxia oculomotor apraxia Type 2 as well as AT and ATLD and it can be difficult to distinguish them at the neurological level. All these disorders, therefore, have cerebellar ataxia as an important feature, but related to AT and ATLD is another disorder, Nijmegen Breakage Syndrome, where there is no cerebellar ataxia. Curiously, the mutated gene here is NBS1. The Nbs1 protein exists in a complex with Mre11 and another protein hRad50 and so the question immediately arises why the clinical features of NBS and ATLD are not more similar. Finally, in this introduction, since ATLD has been identified as a disorder quite recently one can ask whether other similar disorders remain to be identified. I believe that the answer must be an unqualified "yes".

Ataxia Telangiecatasia

AT is inherited in an autosomal recessive manner and the birth frequency in the UK is approximately 1 in 300,000[4] with 5-6 new cases of AT per annum in the UK.[5] There is a wide range of survival in AT, part of the heterogeneity described below, although median survival has been reported in one study as 19 and 25 years.[6] The most frequent causes of death are sinopulmonary infection and malignant disease.[7] These authors suggested that the clinical sine qua non for a diagnosis of ataxia telangiectasia was the presence of both cerebellar ataxia and telangiectasia, seen most strikingly on the bulbar conjunctiva but also present on different parts of the skin.

*A.M.R.Taylor—CR-UK Institute for Cancer Studies, University of Birmingham, Vincent Drive, Birmingham B15 2TT. England. Email: a.m.r.taylor@bham.ac.uk

Molecular Mechanisms of Ataxia Telangiectasia, edited by Shamim I. Ahmad.

Neurology

For most ataxia telangiectasia patients a clear and unambiguous progressive neurological disorder is seen resulting from cerebellar degeneration. This is, by far, the most important clinical feature of AT. Presentation is usually by the age of 2 years, ataxia of both upper and lower limbs develops and by early teenage most patients require a wheelchair for mobility. Movement is affected as shown by the presence of choreoathetosis, myoclonus and tremor. There may be diminished or absent deep reflexes. At the neuropathological level there is loss of Purkinje and granular layer cells.[7] In describing 28 complete autopsy reports, these authors drew attention to a primary cerebellar cortical degeneration and at older ages an additional pathology affecting the cerebrum, brain stem, spinal column and peripheral nerves. At the clinical level with increasing age, into early adulthood, the neurodegeneration affecting the central and peripheral nervous system is shown by loss of vibration sense and position resulting from peripheral nerve damage. Dysarthric speech is also a feature of AT. The eye movement disorder in ataxia telangiectasia is very complex and has been much studied.[8,9] The most important findings in AT include strabismus, apraxia of horizontal gaze, hypometric saccades, pursuit abnormalities and nystagmus.[8] In addition the ocular findings become more prominent with age. A number of measurable neurological traits, including standing and sitting ability, gait head movements, impairment of eye movement, degree of peripheral neuropathy, have been described and shown to be similar between affected AT siblings within families but, most interestingly, different between families where children of the same age are considered. The clear inference is that other genetic or environmental influences may also contribute to the variation in neurological presentation.[10]

Immunodeficiency

Ataxia telangiectasia is a complex disorder that also results in an immunodeficiency. The proportion of individuals with immunodeficiency is variable, probably depending on the population studied and the immunological parameters used. The resulting predisposition to infection is also very variable between patients with most not noticeably affected but others showing frequent episodes of severe infection[11] and approximately 10-15% of patients requiring immunoglobulin replacement. The susceptibility to upper respiratory tract infection is rarely progressive and there is no susceptibility to opportunistic infections.[11] In contrast there is an increase in lower respiratory tract infections with age.[11] This may be partly a consequence of the aspiration of food as a result of impaired swallowing and deficient cough reflex.[12] Recurrent pulmonary infection may lead to chronic lung damage further increasing susceptibility to new infections. We recently published a study of 80 AT patients, where we identified the ATM mutations and determined whether or not the patients had total loss of ATM kinase activity or retained some activity. We were able to show that there was a significantly higher proportion of patients with recurrent sinopulmonary infections in the group of AT patients with mutations resulting in total loss of ATM kinase activity compared with those patients with some residual kinase activity. For two of the most common immunological markers of immunodeficiency in AT patients, IgA deficiency and both T- and B-cell lymphopaenia, patients with abnormalities were found exclusively in the group of patients with no ATM kinase activity. The need for Ig replacement also occurred solely in the group with no ATM kinase activity.[13] This study showed that a there is a strong correlation between the total absence of ATM kinase activity and immunodeficiency in AT and further, that a relatively low level of retained ATM kinase activity is sufficient to allow development of normal immunological function.

Other features of the immunodeficiency in AT include low thymic output, which might explain the T-cell lymphopaenia.[14] Consistent with the fact that the ATM protein is involved in the cellular response to DNA damage and particularly to DNA double strand breaks, the ATM protein has been reported to be involved in stabilizing DNA DSB during V(D)J recombination.[15] In keeping with this, reduced antigen receptor diversity and oligoclonal expansions have been reported in AT patients.[14] There are, however, many unresolved questions concerning the immunodeficiency in AT principal amongst these being why in the complete absence of ATM kinase activity not all patients show clinical immunodeficiency.[13] The involvement of ATM in the lymphoid system is more profound still. More important than the immune deficiency in AT is the predisposition of these patients to both B- and T-cell tumours often in cells with chromosome translocations involving an immune system gene and oncogene.[16] Loss of ATM, therefore, promotes the formation or retention of chromosome translocations predisposing principally to lymphoid malignancies.

Other Features of Ataxia Telangiectasia

A very useful aid in the confirmation of the clinical diagnosis is the consistently found increased level of serum AFP observed in these patients.[17] Until recently, this was the only inherited disorder showing this although recently and quite curiously patients with ataxia oculomotor apraxia Type 2 (AOA2) have also been shown to have an elevated level of serum AFP.[18-21] Some AT patients may show some features of premature ageing, particularly of the skin and including telangiectasia and a papery and atrophied appearance. AT patients may also show growth retardation, which can be quite striking, but it is not a universal feature and may be related to the causes of heterogeneity described below. Endocrine abnormalities and hypogonadism in both sexes are also common features of AT. Menstruation may be delayed and irregular. Hypoplasia of the ovaries had been found at autopsy.[7] However, there have been reports of rare AT patients having children.[22]

Clinical Heterogeneity

The classical, most severe form of AT results from the biallelic total loss of ATM protein kinase activity and loss of ATM function. The cause of this is the presence of two truncating ATM mutations and results in onset of the disorder in early childhood and steady progression through to early adulthood. Experience has shown, however, that there is a marked degree of clinical heterogeneity in AT to the point where each of the two sine qua non diagnostic features, ataxia and telangiectasia, can be very minimally represented even in older affected individuals; this, in the presence of two pathogenic ATM mutations. Indeed every clinical feature of AT, the degree of neurodegeneration, the level of immunodeficiency, the severity of the telangiectasia, can vary between individuals of the same age. What is the cause of this variation? The single largest cause, of the greatest variation, is determined by the type of ATM mutation that is present.

The cause of the clinical heterogeneity is broadly through the presence of one of two types of mutation. First, the presence of a leaky splice site mutation that allows expression of a low level of normal ATM protein that clearly will have normal ATM kinase activity, or a missense mutation that allows the expression of some mutant ATM protein with some residual kinase activity. At the clinical level the consequence of the presence of such mutations can vary from a slightly later age of onset through to onset in early adulthood. Alternatively, onset may be early but progress very slowly. An example of the splice site mutation type is the IVS 40-1050A > G. This substitution creates a splice donor signal that preferentially splices to the downstream exon, which then forces the upstream exon to splice to a cryptic splice acceptor site in the intron, resulting in a 137bp intronic insertion into the coding sequence that causes a frameshift truncating mutation. However this is a leaky mutation and part of the time a normal transcript and hence normal ATM protein is expressed.[23,24] The IVS40 mutation in the homozygous state has been seen in AT patients and this resulted in onset of AT in adulthood[25] compared with a usual age of onset of ~2 years. Another example of a leaky splice site mutation is IVS10-6T > G resulting in expression of some normal ATM and consequently a milder form of AT.[26]

Missense mutations may express mutant ATM protein with residual kinase activity resulting in a milder progression of the neurodegeneration. The 7271T > G (V2424G) mutation has been identified in several families in the UK.[22,27] The V2424G protein is expressed at an almost normal level and although the kinase activity is greatly reduced compared with normal it is still sufficient to give a milder form of AT particularly in those homozygous for the mutation. Similarly a patient, homozygous for the 590G > A (197G > Q) missense mutation, allows expression of mutant ATM with good kinase activity. At the clinical level the patient was able to walk without ataxia at the age of 19 years.[28] Both of these types (splicing and missense) of mutation can result in a very marked amelioration of the clinical phenotype and probably account for the large variation in clinical presentation. These types of mutation are relatively frequently seen; indeed in the UK, the majority of longer lived AT patients have the IVS40-1050A > G mutation. There are probably more individual missense mutations resulting in a milder phenotype than splice site mutations, although more patients may have the latter type of mutation because of a founder effect. Different missense mutations will produce different levels of mutant ATM with different kinase levels and hence variation in clinical phenotype.

A much more rare source of genetic variation resulting in a strikingly milder clinical phenotype is the effect of a possible single modifying gene. This may be exemplified in two siblings reported[29] who while showing no ATM protein at all and homozygosity for a truncating mutation, showed a milder AT. This is an important observation because it infers the presence of a protein that can compensate to some extent for loss of ATM. It is likely that there will be other rare patients like this. These authors also draw attention to the fact that AT mice do not have any overt problem with ataxia. In addition to what, in this case, may be the large effect of a single modifying gene, there are likely to be several if not many other modifying genes although operating at a more subtle level.[29] An example of this is the variation in the individual components, between families but not within families, that constitute the total neurological score for the degree of neurogeneration in AT.[10]

Yet another cause of the heterogeneity of AT is mutation of a different gene to give a phenocopy of AT. The most obvious example of this is ATLD, but also other neurological disorders such as ataxia oculomotor apraxia 1 (AOA1) and AOA2 (see below).

Ataxia Telangiectasia Like Disorder (ATLD) and Nijmegen Breakage Syndrome (NBS)

Ataxia telangiectasia like disorder (ATLD) is very rare, with only 16 cases published, four from the UK,[30-32] two from Italy[33] and ten from Saudi Arabia.[34] The clinical features of ATLD are very similar to those of AT, principally the progressive cerebellar ataxia, although onset is at a slightly later age of childhood and slower in progress; indeed in two cases a diagnosis of benign hereditary chorea was initially made.[35] There is the appearance, therefore, of a milder form of AT. Unlike AT, ATLD patients do not show the presence of telangiectasia. ATLD patients also show normal levels of total IgM, IgA and IgG, although there may be reduced levels of specific functional antibodies. hMre11 is a component of a complex containing the Nbs1 protein. It might be expected therefore, that deficiency of hMre11 would result in a clinical phenotype more like Nijmegen Breakage Syndrome where there is deficiency of Nbs1 protein.[36-38] Why this is not the case is not understood. In contrast to AT and ATLD, NBS patients characteristically show microcephaly, some learning difficulties, immunodeficiency and a greater predisposition to lymphoid trumours than AT patients. There is no cerebellar degeneration. While AT, ATLD and NBS can all be distinguished at the clinical level, they all show increased levels of chromosome abnormalities involving chromosomes 7 and 14. At the cellular level all three disorders exhibit hypersensitivity to ionising radiation—the cellular phenotype is therefore similar. The function of the Mre11 complex is required for full ATM activation and so loss of Mre11 might explain the ATLD phenotype.

It is not known whether any ATLD patients have a predisposition to cancer as too few patients have been described so far although none, so far, has developed a tumour and most are adults.This is in contrast to AT where lymphoid tumours in particular are a feature of childhood. An important feature of both AT and NBS patients is their very high risk for developing cancer, principally lymphoid tumours[16,39] An Mre11 mouse model has been derived[40] with the truncating mutation of hMRE11 ATLD patient 1[31] but showed no development of tumours. This is particularly interesting since Nijmegen Breakage Syndrome patients with mutations in NBS1 (another component of the MRN complex) show possibly a greater predisposition to lymphoid tumours than AT patients.

Unlike ATM, hMRE11, NBS1 and RAD50 are all essential genes; total loss of hMRE11, like NBS1 and hRAD50 is lethal. The mutations in hMRE11 giving ATLD are, therefore, hypomorphic mutations allowing expression of either truncated or full length mutant protein, with some retained function.[24] It is not clear from the limited numbers of patients identified how prevalent clinical heterogeneity might be within ATLD. It is notable, however, that the patients described[34] had additional features, particularly microcephaly (more associated with ~Nijmegen Breakage Syndrome). These patients were homozygous for an MRE11 missense mutation. In contrast, NBS is genetically very uniform, the vast majority of patients having the 657del5 NBS1 mutation and, therefore, there is little scope for clinical heterogeneity of the magnitude mediated by the different mutations in AT, although some variations have been reported associated with mutations different to the 657del5.[42,43]

Conclusions

The recent report of the RIDDLE syndrome (Stewart et al, 2007), with some overlapping features of ataxia telangiectasia shows that other clinical phenotypes are yet to be identified, caused by mutation of other genes but where there is similarity of the cellular phenotype and involvement in the DNA damage response pathway as in AT, ATLD and NBS. The identification of a wide clinical heterogeneity in AT makes it almost certain that additional clinical phenotypes will be discovered. Understanding the origin of these phenotypes together with the possibility of genes compensating for ATM will improve our understanding of this disorder the precise roles of the ATM protein in neurodegeneration and the prospects for treatment of this serious disorder.

Acknowledgements

I thank CR-UK and The Ataxia Telangiectasia Society of the UK for continued support.

References

1. Aicardi J, Barbosa C, Andermann E et al. Ataxia-ocular motor apraxia: a syndrome mimicking ataxia telangiectasia. Ann Neurol 1988; 24:497-502.
2 Moreira MC, Barbot C, Tachi N et al. The gene mutated in ataxia-ocular apraxia 1 encodes the new HIT/Zn-finger protein aprataxin. Nat Genet 2001; 29:189-193.
3. Date H, Onodera O, Tanaka H et al. Early-onset ataxia with ocular motor apraxia and hypoalbuminemia is caused by mutations in a new HIT superfamily gene. Nat Genet 2001; 29:184-188.
4. Woods CG, Bundey SE, Taylor AM. Unusual features in the inheritance of ataxia telangiectasia. Hum Genet 1990; 84:555-562.
5. Thompson D, Duedal S, Kirner J et al. Cancer risks and mortality in heterozygous ATM mutation carriers. J Natl Cancer Inst 2005; 97(11):813-822.
6. Crawford TO, Skolasky RL, Fernandez R et al. Survival probability in ataxia telangiectasia. Arch Dis Child 2006; 91:610-611.
7. Sedgwick RP, Boder E. Ataxia Telangiectasia. In: de Jong JBMV, ed. Handbook of Clinical Neurology. Hereditary Neuropathies and Spinocerebellar Atrophies. Amsterdam: Elsevier Science Publishers BV 1991:347-352.
8 Farr AK, Shalev B, Crawford TO et al. Ocular manifestations of ataxia telangiectasia. Am J Ophthalmol 2002; 134:891-906.
9. Lewis RF, Crawford TO. Slow target-directed eye movements in ataxia telangiectasia. Invest Ophthalmol Vis Sci 2002; 43:686-691.
10. Crawford TO, Mandir AS, Lefton-Greif MA et al. Quantitative neurologic assessment of ataxia telangiectasia. Neurology 2000; 54:1505-1509.
11. Nowak-Wegrzyn A, Crawford TO, Winkelstein JA et al. Immunodeficiency and infections in ataxia telangiectasia. J Pediatr 2004; 144(4):505-511.
12. Lefton-Greif MA, Crawford TO, Winkelstein JA et al. Oropharyngeal dysphagia and aspiration in patients with ataxia telangiectasia. J Pediatr 2000; 136:225-231.
13. Staples ER, McDermott EM, Reiman A et al. Immunodeficiency in ataxia telangiectasia is correlated strongly with the presence of two null mutations in the ataxia telangiectasia mutated gene. Clin Exp Immunol 2008; 153:214-220.
14. Giovannetti A, Mazzetta F, Caprini E et al. T-cell receptor repertoire, decreased thymic output and predominance of terminally differentiated T-cells in ataxia telangiectasia. Blood 2002; 100:4082-4089.
15. Bredemeyer AL, Sharma GG, Huang CY et al. ATM stabilizes DNA double-strand-break complexes during V(D)J recombination. Nature 2006; 442:466-470.
16. Taylor AM, Metcalfe JA, Thick J et al. Leukemia and lymphoma in ataxia telangiectasia. Blood 1996; 87:423-438.
17. Waldmann TA, McIntire KR. Serum alpha-feto-protein levels in patients with ataxia telangiectasia. Lancet 1987; 25:1112-1115.

18. Moreira MC, Klur S, Watanabe M et al. Senataxin, the ortholog of a yeast RNA helicase, is mutant in ataxia-ocular apraxia 2. Nat Genet 2004; 36(3):225-227.
19. Le Ber I, Bouslam N, Rivaud-Pechoux S et al. Frequency and phenotypic spectrum of ataxia with oculomotor apraxia 2: a clinical and genetic study in 18 patients. Brain 2004; 127:759-767.
20. Nemeth AH, Bochukova E, Dunne E et al. Autosomal recessive cerebellar ataxia with oculomotor apraxia (ataxia telangiectasia-like syndrome) is linked to chromosome 9q34. Am J Hum Genet 2000; 67:1320-1326.
21. Izatt L, Nemeth AH, Meesaq A et al. Autosomal recessive spinocerebellar ataxia and peripheral neuropathy with raised alpha-fetoprotein. J Neurol 2004; 251:805-812.
22. Stankovic T, Kidd AM, Sutcliffe A et al. ATM mutations and phenotypes in ataxia telangiectasia families in the British Isles: expression of mutant ATM and the risk of leukemia, lymphoma and breast cancer. Am J Hum Genet 1998; 62:334-345.
23. McConville CM, Stankovic T, Byrd PJ et al. Mutations associated with variant phenotypes in ataxia telangiectasia. Am J Hum Genet 1996; 59:320-330
24. Stewart GS, Last JI, Stankovic T et al. Residual ataxia telangiectasia mutated protein function in cells from ataxia telangiectasia patients, with 5762ins137 and 7271T→G mutations, showing a less severe phenotype. J Biol Chem 2001; 276:30133-30141.
25. Sutton IJ, Last JI, Ritchie SJ et al. Adult-onset ataxia telangiectasia due to ATM 5762ins137 mutation homozygosity. Ann Neurol 2004; 55:891-895.
26. Austen B, Barone G, Reiman A et al. Pathogenic ATM mutations occur at a low frequency in multiple myeloma. Brit J Haematol; 142:925-933.
27. Taylor AM. Byrd PJ. Molecular pathology of ataxia telangiectasia. J Clin Pathol 2005; 58:1009-1015.
28. Carrillo F, Schneider SA, Taylor AMR et al. Prominent oromandibular dystonia and pharyngeal telangiectasia in atypical ataxia telangiectasia. Cerebellum 2009; 8(1):22-27.
29. Alterman N, Fattal-Valevski A, Moyal L et al. Ataxia telangiectasia: mild neurological presentation despite null ATM mutation and severe cellular phenotype. Am J Med Genet 2007; PartA143A:1827-1834.
30. Hernandez D, McConville CM, Stacey M et al. A family showing no evidence of linkage between the ataxia telangiectasia gene and chromosome 11q22-23. J Med Genet 1993; 30:135-140.
31. Stewart GS, Maser RS, Stankovic et al. The DNA double-strand break repair gene hMRE11 is mutated in individuals with an ataxia telangiectasia-like disorder. Cell 1999; 99:577-587.
32. Pitts SA, Kullar HS, Stankovic T et al. HMRE11: genomic structure and a null mutation identified in a transcript protected from nonsense-mediated mRNA decay. Hum Mol Genet 2001; 10:1155-1162.
33. Delia D, Piane M, Buscemi G et al. MRE11 mutations and impaired ATM-dependent responses in an Italian family with ataxia telangiectasia-like disorder. Hum Mol Genet 2004; 13:2155-2163.
34. Fernet M, Gribaa M, Salih MA et al. Identification and functional consequences of a novel MRE11 mutation affecting ten Saudi Arabian patients with the Ataxia telangiectasia-like disorder (ATLD). Hum Mol Genet 2005; 14(2):307-318.
35. Klein C, Wenning GK, Quinn NP et al. Ataxia without telangiectasia masquerading as benign hereditary chorea. Mov Disord 1996; 11:217-220.
36. Taalman RD, Jaspers NG, Scheres JM et al. Hypersensitivity to ionizing radiation, in vitro, in a new chromosomal breakage disorder, the Nijmegen Breakage Syndrome. Mutat Res 1983; 112:23-1132.
37. Varon R, Vissinga C, Platzer M et al. Nibrin, a novel DNA double-strand break repair protein, is mutated in Nijmegen breakage syndrome. Cell 1998; 93:467-476.
38. Carney JP, Maser RS, Olivares H et al. The hMre11/hRad50 protein complex and Nijmegen breakage syndrome: linkage of double-strand break repair to the cellular DNA damage response. Cell 1998; 93:477-486.
39. International Nijmegen Breakage Study Group. Nijmegen breakage syndrome. The International Nijmegen Breakage Syndrome Study Group. Arch Dis Child 2000; 82:400-406.
40. Theunissen JW, Kaplan MI, Hunt PA et al. Checkpoint failure and chromosomal instability without lymphomagenesis in Mre11(ATLD1/ATLD1) mice. Mol Cell 2003; 12:1511-1523.
41. Varon R, Dutrannoy V, Weikert G et al. Nijmegen breakage syndrome phenotype due to alternative splicing. Hum Mol Genet 2006; 15(5):679-689.
42. Seemanová E, Sperling K, Neitzel H et al. Nijmegen breakage syndrome (NBS) with neurological abnormalities and without chromosomal instability. Med Genet 2006; 43(3):218-224.
43. Stewart GS, Stankovic T, Byrd PJ et al. RIDDLE immunodeficiency syndrome is linked to defects in 53BP1-mediated DNA damage signaling. Proc Natl Acad Sci USA 2007; 104:16910-16915.

Chapter 2

Mutations in the Ataxia Telangiectasia Mutated (ATM) Gene

Akira Tachibana*

Abstract

Ataxia telangiectasia (AT) is a multisystem autosomal recessive disorder by mutations in a single gene, *ATM*. A large number of mutations in this gene have been identified so far. In most cases of AT patients, null *ATM* mutations lead to truncated or greatly destabilized protein. AT variants, who show milder manifestations of the clinical or cellular characteristics of the disease, exhibit reduced but detectable level of the ATM protein; these patients show the expected mild phenotype. The large size of the *ATM* gene and the diversity and broad distribution of mutations in AT patients limit the efficient screening of mutations as a diagnostic tool.

Introduction

Ataxia telangiectasia (AT) is an autosomal recessive disorder characterized by cerebellar degeneration, immunodeficiency, chromosomal instability, radiation sensitivity and cancer predisposition. Most prominent are the sensitivity of cells from patients to ionizing radiation and radiomimetic chemicals and defect in the activation of cell-cycle checkpoints after treatment with these agents. The responsible gene is *ATM*, which encodes a large protein kinase with a phosphatidylinositol 3-kinase-like domain.[1,2] The diverse description of this disease are indications of the pleiotropic effects, that the *ATM* gene exerts.

In order to clarify the relationships between genotype and phenotype of AT, analysis of mutation in the responsible gene is very important. In this chapter, ATM gene and characteristics of the mutations in the gene are described in relation to the disease. Although hundreds of different pathogenic mutations have been described to date, only selected ones will be discussed here.

Complementation Groups

Genetic heterogeneity has been observed as complementation groups in some hereditary diseases with DNA repair deficiency, such as xeroderma pigmentosum (XP) and Fanconi anemia (FA). For XP and FA, identification of complementation groups has been very useful, since each complementation group corresponds to a single gene responsible for the disease.[3,4] To identify complementation groups in AT patients, attempts have been made since early 1980s.[5-7] The radioresistant DNA synthesis was used for the basis of detection of the complementation in heterokaryons obtained by fusing cells from different patients. These studies consistently represented phenotype complementation between certain patients and identified four complementation groups, designated groups A/B, C, D and E, of which A/B is the largest group.

However, the identification of the *ATM* gene as a single gene, responsible for AT, ruled out that the apparent complementation did not reflect the true genetic heterogeniety.[8] The reason

*Akira Tachibana—Faculty of Science, Ibaraki University, 2-1-1, Bunkyo, Mito, Ibaraki 310-8512, Japan. Email: akiratac@mx.ibaraki.ac.jp

Molecular Mechanisms of Ataxia Telangiectasia, edited by Shamim I. Ahmad.

why the multiple complementation groups had been identified despite a single gene is not yet fully understood, but intragenic complementation is probable. Another possibility is that the cellular phenotype used in these studies, radioresistant DNA synthesis, may not reflect the direct effect of the primary genetic defects in this disorder.

AT Variants

AT patients have been grouped together and have not been subdivided into clinical subtypes, since its diagnosis is usually unambiguous in most of the AT patients, despite some variability in several features of the disease. Although the majority of AT patients show the uniform phenotype, milder cases of the disease have occasionally been reported. These rare cases of AT patients with milder manifestations of the clinical or cellular characteristics of the disease have been designated "AT variants".[1,9] Such patients show onset of the disease either at later age, moderation in severity of the ataxia and the progression of the disease is slower. Also shows intermediate chromosomal instability, cellular radiosensitivity and longer life span than is seen in most AT patients. Some of these patients may not have telangiectasia, but their disease may still result from *ATM* mutations. These patients who show the typical AT clinical features are designated "classical AT patients".

Identification of the *ATM* Gene

Savitsky et al[8] identified the *ATM* gene by positional cloning and they found mutations in the gene from the AT patients belonging to all four complementation groups. This suggested that the *ATM* gene is a single gene responsible for AT and that the complementation groups reported previously did not reflect the true genetic heterogeneity. Hereafter, "AT patients" are defined as those patients whose *ATM* genes are suffered from certain mutations.

Mutations in Classical AT Patients

Savitsky et al[8] also analyzed cDNA of the *ATM* gene from 14 AT patients and identified ten mutations. One of them was a large deletion that lost most of open reading frame. Six were small deletions or insertions, which caused frameshift mutations and predicted to result in truncated proteins. Three of them were in-frame deletions resulting in deletions of one, two, three amino acids, respectively. Thus, seven of ten mutations were supposed to resulting in truncated proteins which would cause significant effect on the gene product. All the mutations identified in the *ATM* cDNA sequence were either deletions or insertions and no base substitutions were identified.

Soon after this report, many studies have been carried out to clarify the nature of the *ATM* mutations.[10-29] These sequence analyses revealed that mutations that would truncate the protein were predominant among classical AT patients. For example, Gilad et al[13] reported that 39 of 44 mutations identified were either incorrect initiations or terminations of translation, or deletions of large segments from the gene product. These irregular gene products, most of which were shorter forms, were predicted to be unstable leading to no ATM protein in the patients. This assumption was confirmed by the analyses which identified no ATM protein in classical AT patients. These results indicate that the classical AT phenotype results from null alleles of *ATM* and the activity of ATM protein is totally abolished in patients' cells. Most of the mutations that cause truncation were small deletions or insertions of one to several bases, which would result in frameshift, but there were some small deletions that caused in-frame deletion resulting in removal of several amino acids from the gene product. Four base substitutions that changed a normal codon to a termination codon were also identified. On the other hand, one base substitution mutation that changed the termination codon of the *ATM* gene to serine codon was detected. This mutation would result in extension of protein by 29 amino acids. Two mutations (one deletion and one base substitution) affected the initiation codon to abolish the initiation of protein synthesis. Some deletions that remove hundreds of bases from the cDNA sequence were found and most of them were the results of skipping of one or two exons. While the majority of the *ATM* mutations are those that would affect the protein structure, some missense mutations have also been found in AT patients.

Certain missense mutations found in AT patients seem to destabilize the protein, with the same consequence as the mutations truncating the protein.[30,31]

One in-frame deletion *7636del9*, which removed 9 base pairs from exon 53, was identified in a number of AT families and suggested that this was one of the most frequent variant of the *ATM* gene.[16] The mutation has also been identified repeatedly in human malignant cells (see "Mutations at the *ATM* gene and malignancy").

In summary, most AT patients are compound heterozygotes and the mutations are broadly distributed throughout the gene. A wide variety of mutations have been identified in AT patients and most of them are unique to single families. Mutations that cause protein truncations are predominant among the classical AT patients, which result in the absence of the functional protein in AT patients. This feature of *ATM* mutations is supposed to be the basis for the uniform nature of clinical phenotype among classical AT patients.

Mutations in AT Variant Patients

As described above, AT variants show milder manifestations of clinical and cellular features of the disease. To clarify the molecular nature for the variation of the phenotype, the *ATM* gene was investigated in AT variants.[32-34] McConville et al[32] analyzed 14 AT variant families and found that 10 of them have a 137-bp insertion in their cDNA (*5762ins137*), which was predicted to cause the truncation of the ATM protein. Genomic DNA sequence analysis revealed that a base substitution of A > G in intron 40, that activated a cryptic splice donor site, resulted in a 137-bp insertion of intronic sequence in the transcript. Further studies of 2 brothers who were homozygous for this mutation revealed that, besides the aberrant transcript with a 137-bp insertion, normal *ATM* transcript was also produced from this mutant allele, although the level of the normal transcript was significantly reduced.[35] The result suggested that this mutation is a leaky splicing mutation. McConville et al[32] also identified 2 base substitutions, *7271T > G* and *8480T > G*, which resulted in amino acid changes, Val2424Gly and Phe2827Cys, respectively.

Gilad et al[33] analyzed 6 Italian AT variant patients and found a mutation, *3576G > A*, at the splicing donor site of exon 26 in 4 patients. This mutation caused the skipping of exon 26, leading to 174-bp deletion in the transcript, resulting in in-frame deletion of 58 amino acids in the gene product. However, this mutation is likely to be a leaky splicing mutation, as two patients who were homozygous for this mutant allele showed residual amount of ATM protein.[33]

The fact that all the AT variants exhibited reduced but detectable level of ATM protein is a clear contrast to that undetectable level of ATM protein in classical AT patients. Thus, these leaky splicing mutations are considered to contribute to the pathogenesis of AT variants.

Founder Mutations

As shown above, most *ATM* mutations are unique and distributed throughout the gene. However, some ethnic characteristics have been observed and considered as founder effects.

A single mutation, *103C > T*, was found in 32 of 33 defective *ATM* alleles in a total of 17 Jewish North African AT families.[12] This mutation results in a termination codon at position 35 of the ATM protein.

A complex mutation *3245ATC > TGAT* at codon 1082, which caused a termination at codon 1095, was found in 12 of 22 defective *ATM* alleles in 11 Norwegian AT families; 5 homozygous patients and 2 compound heterozygous patients.[22] Genealogical investigations revealed that a woman born in 1684 was a common ancestor for three of the families.

The analysis of 8 unrelated Japanese AT families[19] revealed that 44% (7 out of 16) of the mutant alleles had one of two mutations, *7884del5* and *IVS33 + 2T > C*. Microsatellite genotyping, surrounding the *ATM* locus, indicated that a common haplotype was shared by the mutant alleles in both mutations, suggesting that these two founder mutations are predominant among the Japanese ATM mutant alleles. The latter mutation, *IVS33 + 2T > C*, will be discussed below. The prevalence of the two mutations in Japanese patients was confirmed by another analysis of mutations in 4 Japanese AT families.[20]

AT patients in 68 families in the British Isles were analyzed and 59 mutations were identified.[24] Eleven of 51 mutations in 60 families, native to the British Isles, were identified in more than one family and confirmed as founder mutations by the presence of a common haplotype within the families. Two of the 11 founder mutations resulted in a milder clinical phenotype, one of which was *5762ins137* that has already been discussed in the section "Mutations in AT variant patients".

Analysis of mutations in 24 Polish AT families[28] identified 3 founder mutations, which accounted for 58% of the mutant alleles. Three-quarters of the AT families had at least one recurring mutation notwithstanding the low frequency of consanguinity in Poland.

Mutations at the *ATM* Gene and Malignancy

AT patients have a predisposition to malignancy, especially to particular types of leukemia and lymphoma.[36] In particular, T-prolymphocytic leukemia (T-PLL), a rare and sporadic leukemia, has similarities to a mature T-cell leukemia seen in some AT patients. In addition, it has been estimated that female relatives of AT patients have an excess risk of breast cancer.[37] Therefore, analyses of mutations at the *ATM* gene in patients of a large number of leukemia and lymphoma, breast cancer and other cancers have been extensively carried out.[38-52] The studies have identified quite a few *ATM* mutations in malignant cells. The majority of mutations found in malignant cells including T-PLL, B-cell non-Hodgkin's lymphoma, mantle cell lymphoma, breast cancer and kidney cancer, were base substitutions resulting in a change of one amino acid of the protein. Below only some important data are discussed.

The *7271T > G* mutation, identified in AT variants, that results in an amino acid substitution Val2424Gly,[32] was found in T-PLL tissue[41] and in breast cancer.[52] In a population-based study, the heterozygous 7271T > G mutation was identified in 0.2% (7 of 3743) of breast cancer patients but none of 1268 controls.[53] The estimated cumulative breast cancer risk to age 70 years was 52%. Another mutation *IVS10-6T > G* has also been suggested to be susceptible to breast cancer.[52] This mutation causes the skipping of exon 11, resulting in a frameshift and subsequent truncation of the protein at codon 419. However, further analyses[53,54] using large number of families suggested that *IVS10-6T > G* mutation is not associated with breast cancer and that *7271T > G* mutation is associated with a substantially increased breast cancer risk.

The mutation *7636del9*, which has been frequently identified in AT patients[16] (see "Mutations in classical AT patients"), was also found in T-PLL patients[41] and in a female breast cancer patient with a family history of multiple malignancies.[39] It was demonstrated that *Atm* knock-in heterozygous mice containing an in-frame deletion, corresponding to this mutation had an increased susceptibility to tumors.[55] This result is a clear contrast to *Atm* knockout heterozygous mice ($Atm^{+/-}$) that showed no tumors. Expression of the cDNA containing this mutation demonstrated a dominant-negative effect in cells, that is, the ATM kinase activity induced by radiation is inhibited by this mutation both in vivo and in vitro.

Characteristic *ATM* Mutations

AT_{Fresno} *(IVS33 + 2T > C)*

A special variant form of AT was identified in identical twin girls, which combines a classical AT phenotype with microcephaly and mental retardation and was designated "AT_{Fresno}".[56] Analysis of its *ATM* gene[33] revealed that the pathogenic mutation was IVS33 + 2T > C, which is a typical AT mutation. This mutation abolished a splice site at intron 33, leading to skipping of exon 32. This caused a deletion of 165 nucleotides resulting in in-frame deletion of 55 amino acids. There was no immunoreactive ATM protein in the patient's cell,[33] which suggests that the gene product with this mutation is unstable. As described above, this mutation was indicated as one of two founder mutations among the Japanese alleles.[19,20]

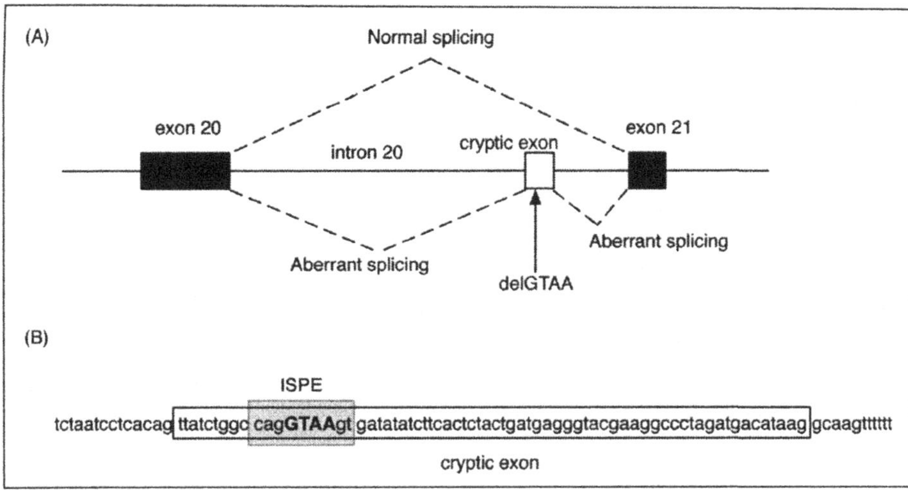

Figure 1. The effect of IVS20-579delGTAA on splicing. A) Schematic representation of the splicing of intron 20 in the *ATM* gene. Authentic exons are indicated as filled boxes, introns as line and dotted lines represent possible splicing products. An open box shows the cryptic exon. The arrow indicates the approximate position of the mutation. B) Part of the sequence of intron 20 including the cryptic exon is represented. The cryptic exon is shown by open box and ISPE by shadowed box. The 4 nucleotides deleted in intron 20 are indicated with uppercase letters.

IVS20-579delGTAA (4-bp del, 65-bp ins)

An unusual type of insertion mutation was identified,[57] which includes a cryptic exon of 65 bp with a deletion of 4 nucleotides (GTAA) in intron 20. Analysis of splicing defect identified a new intron-splicing processing element (ISPE) (Fig. 1). This element is complementary to U1 snRNA that is a component of the U1 small nuclear ribonucleoprotein (snRNP) and mediates accurate intron processing by interacting with U1 snRNP. Further studies[58] demonstrated that U1 snRNP binding at the ISPE interferes with the cryptic acceptor site and that activation of this site leads to an accurate 5′-3′ order of intron removal around the cryptic exon. A deletion of GTAA within ISPE abolishes the interaction with U1 snRNP and activates the cryptic exon.

This type of mutation has been designated as "pseudoexon insertion" and classified as Type II nonclassical splicing mutation comparing to classical splicing mutation (Type I). It has been shown that four patients of various ethnicity with this mutation, shared a common founder haplotype.[59] It has also been pointed out that, although some mutations had been interpreted as classical splicing mutations, they were in fact nonclassical splicing mutations, which would produce different types of *ATM* transcripts leading to distinct gene products. This caused further misinterpretation of the genotype-phenotype relationships.

Concluding Remarks

The large size of the *ATM* gene with 66 exons, spanning approximately 150 kb of genomic DNA and the diversity and broad distribution of mutations in AT patients, makes direct mutation screening difficult. Also mistakes in the splicing process may have significant impact in the gene products. In addition, a mutation in intron that affect splicing revealed a new regulatory mechanism of splicing.[57,59] These new findings of nonclassical splicing mutations made us re-evaluate the interpretations for mutations. Particularly these genomic mutations without examining cDNA might cause misinterpretation of the relationships between genotype and phenotype, which could have certain impact on procedures for therapy. Therefore it should be useful to develop novel methods to identify mutations easily and efficiently in the *ATM* gene for diagnosis and carrier identifications.

References

1. Shiloh Y. Ataxia telangiectasia and the Nijmegen breakage syndrome: Related disorders but genes apart. Annu Rev Genet 1997; 31:635-662.
2. Lavin MF, Shiloh Y. The genetic defect in ataxia telangiectasia. Annu Rev Immunol 1997; 15:177-222.
3. De Boer J, Hoeijmakers JH. Nucleotide excision repair and human syndromes. Carcinogenesis 2000; 21:453-460.
4. Patel KJ, Joenje H. Fanconi anemia and DNA replication repair. DNA Repair 2007; 6:885-890.
5. Jaspers NG, Bootsma D. Genetic heterogeneity in ataxia telangiectasia studied by cell fusion. Proc Natl Acad Sci USA 1982; 79:2641-2644.
6. Murnane JP, Painter RB. Complementation of the defects of DNA synthesis in irradiated and uiirradiated ataxia telangiectasia cells. Proc Natl Acad Sci USA 1982; 79:1960-1963.
7. Jaspers NG, Gatti RA, Baan C et al. Genetic complementation analysis of ataxia telangiectasia and nijmegen breakage syndrome: a survey of 50 patients. Cytogenet Cell Genet 1988; 49:259-263.
8. Savitsky K, Bar-Shira A, Gilad S et al. A single ataxia telangiectasia gene with a product similar to PI-3 kinase. Science 1995; 268:1749-1753.
9. Sedgwick RP, Boder E. Ataxia telangiectasia. In: de Jong JMBV, ed. Handbook of Clinical Neurology: Hereditary Neuropathies and Spinocerebellar Atrophines. Amsterdam: Elsevier, 1991; 16:347-423.
10. Baumer A, Bernthaler U, Wolz W et al. New mutations in the ataxia telangiectasia gene. Hum Genet 1996; 98:246-249.
11. Byrd PJ, McConville CM, Cooper P et al. Mutations revealed by sequencing the 5′ half of the gene for ataxia telangiectasia. Hum Mol Genet 1996; 5:145-149.
12. Gilad S, Bar-Shira A, Harnik R et al. Ataxia telangiectasia: founder effect among north african jews. Hum Mol Genet 1996; 5:2033-2037.
13. Gilad S, Khosravi R, Shkedy D et al. Predominance of null mutations in ataxia telangiectasia. Hum Mol Genet 1996; 5:433-439.
14. Telatar M, Wang Z, Udar N et al. Ataxia telangiectasia: mutations in ATM cDNA detected by protein-truncation screening. Am J Hum Genet 1996; 59:40-44.
15. Vorechovsky I, Luo L, Prudente S et al. Exon-scanning mutation analysis of the ATM gene in patients with ataxia telangiectasia. Eur J Hum Genet 1996; 4:352-355.
16. Wright J, Teraoka S, Onengut S et al. A high frequency of distinct ATM gene mutations in ataxia telangiectasia. Am J Hum Genet 1996; 59:839-846.
17. Concannon P, Gatti RA. Diversity of ATM gene mutations detected in patients with ataxia telangiectasia. Hum Mutat 1997; 10:100-107.
18. Broeks A, de Klein A, Floore AN et al. ATM germline mutations in classical ataxia telangiectasia patients in the dutch population. Hum Mutat 1998; 12:330-337.
19. Ejima Y, Sasaki MS. Mutations of the ATM gene detected in Japanese ataxia telangiectasia patients: possible preponderance of the two founder mutations 4612del165 and 7883del5. Hum Genet 1998; 102:403-408.
20. Fukao T, Song X, Yoshida T et al. Ataxia telangiectasia in the japanese population: Identification of R1917X, W2491R, R2909G, IVS33+2->A and 7883del5, the latter two being relatively common mutations. Hum Mutat 1998; 12:338-343.
21. Gilad S, Khosravi R, Hamik R et al. Identification of ATM mutations using extended RT-PCR and restriction endonuclease fingerprinting and elucidation of the repertoire of AT mutations in Israel. Hum Mutat 1998; 11:69-75.
22. Laake K, Telatar M, Geitvik GA et al. Identical mutations in 55% of the ATM alleles in 11 norwegian AT families: evidence for a founder effect. Eur J Hum Genet 1998; 6:235-244.
23. Sasaki T, Tien H, Kukita Y et al. ATM mutations in patients with ataxia telangiectasia screened by a hierarchical strategy. Hum Mutat 1998; 12:186-195.
24. Stankovic T, Kidd AMJ, Sutcliffe A et al. ATM mutations and phenotypes in ataxia telangiectasia families in the british isles: expression of mutant ATM and the risk of leukemia, lymphoma and breast cancer. Am J Hum Genet 1998; 62:334-345.
25. Telatar M, Teraoka S, Wang Z et al. Ataxia telangiectasia: identification and detection of founder-effect mutations in the ATM gene in ethnic populations. Am J Hum Genet 1998; 62:86-97.
26. Sandoval N, Platzer M, Rosenthal A et al. Characterization of ATM gene mutations in 66 ataxia telangiectasia families. Hum Mol Genet 1999; 8:69-79.
27. Laake K, Jansen L, Hahnemann JM et al. Characterization of ATM mutations in 41 Nordic families with ataxia telangiectasia. Hum Mutat 2000; 16:232-246.
28. Babaei M, Mitui M, Olson ER et al. ATM haplotypes and associated mutations in Iranian patients with ataxia telangiectasia: recurring homozygosity without a founder haplotype. Hum Genet 2005; 117:101-106.
29. Mitui M, Bernatowska E, Pietrucha B et al. ATM gene founder haplotypes and associated mutations in polish families with ataxia telangiectasia. Ann Hum Genet 2005; 69:657-664.

30. Lakin ND, Weber P, Stankovic T et al. Analysis of the ATM protein in wild type and ataxia telangiectasia cells. Oncogene 1996; 13:2707-2716.
31. Ziv Y, Bar-Shira A, Pecker I et al. Recombinant ATM protein complements the cellular AT phenotype. Oncogene 1997; 15:159-167.
32. McConville CM, Stankovic T, Byrd PJ et al. Mutations associated with variant phenotypes in ataxia telangiectasia. Am J Hum Genet 1996; 59:320-330.
33. Gilad S, Chessa L, Khosravi R et al. Genotype-phenotype relationships in ataxia telangiectasia and variants. Am J Hum Genet 1998; 62:551-561.
34. Dort T, Bendix-Waltes R, Wegner RD et al. Slow progression of ataxia telangiectasia with double missense and in frame splice mutations. Am J Med Genet 2004; 126:272-277.
35. Sutton IJ, Last JI, Ritchie SJ et al. Adult-onset ataxia telangiectasia due to ATM 5762ins137 mutation homozygosity. Ann Neurol 2004; 55:891-895.
36. Taylor AMR, Metcalfe JA, Thick J et al. Leukemia and lymphoma in ataxia telangiectasia. Blood 1996; 87:423-438.
37. Swift M, Reitnauer PJ, Morrell D et al. Breast and other cancers in families with ataxia telangiectasia. N Engl J Med 1987; 316:1289-1294.
38. Vorechovsky I, Luo L, Lindblom A et al. ATM mutations in cancer families. Cancer Res 1996; 56:4130-4133.
39. Vorechovsky I, Rasio D, Luo L et al. The ATM gene and susceptibility to breast cancer: analysis of 38 breast tumors reveals no evidence for mutation. Cancer Res 1996; 56:2726-2732.
40. Stilgenbauer S, Schaffner C, Litterst A et al. Biallelic mutations in the ATM gene in T-prolymphocytic leukemia. Nat Med 1997; 3:1155-1159.
41. Vorechovsky I, Luo L, Dyer MJ et al. Clustering of missense mutations in the ataxia telangiectasia gene in a sporadic T-cell leukaemia. Nat Genet 1997; 17:96-99.
42. Luo L, Lu F, Hart S et al. Ataxia telangiectasia and T-cell leukemia's: no evidence for somatic ATM mutation in sporadic T-ALL or for hypermethylation of the ATM-NPAT/E14 bidirectional promoter in T-PLL. Cancer Res 1998; 58:2293-2297.
43. Yulle MA, Coignet LJ. The ataxia telangiectasia gene in familial and sporadic cancer. Recent Results Cancer Res 1998; 154:156-173.
44. Yullie MAR, Coignet LJA, Abraham SM et al. ATM is usually rearranged in T-cell prolymphocytic leukemia. Oncogene 1998; 16:789-796.
45. Bevan S, Catovsky D, Marossy A et al. Linkage analysis for ATM in familial B-cell chronic lymphocytic leukemia. Leukemia 1999; 13:1497-1500.
46. Schaffner C, Stilgenbauer S, Rappold GA et al. Somatic ATM mutations indicate a pathogenic role of ATM in B-cell chronic lymphocytic leukemia. Blood 1999; 94:748-753.
47. Stilgenbauer S, Winkler D, Ott G et al. Molecular characterization of 11q deletions points to a pathogenic tole of the ATM gene in mantle cell lymphoma. Blood 1999; 94:3262-3264.
48. Broeks A, Urbanus JHM, Floore AN et al. ATM-heterozygous germline mutations contribute to breast cancer-susceptibility. Am J Hum Genet 2000; 66:494-500.
49. Dork T, Bendix R, Bremer M et al. Spectrum of ATM gene mutations in a hospital-based series of unselected breast cancer patients. Cancer Res 2001; 61:7608-7615.
50. Lu Y, Condie A, Bennett JD et al. Disruption of the ATM gene in breast cancer. Cancer Genet Cytogenet 2001; 126:97-101.
51. Yullie MR, Condie A, Hudson CD et al. ATM mutations are rare in familial chronic lymphocytic leukemia. Blood 2002; 100:603-609.
52. Bernstein JL, Bernstein L, Thompson WD et al. ATM variants 7271T > G and IVS10-6T > G among women with unilateral and bilateral breast cancer. Br J Cancer 2003; 89:1513-1516.
53. Bernstein JL, Teraoka S, Southey MC et al. Population-based estimates of breast cancer risks associated with ATM gene variants c.7271T > G and c.1066-6T > G (IVS10-6T > G) from the breast cancer family registry. Hum Mutat 2006; 27:1122-1128.
54. Szabo CI, Schutte M, Broeks A et al. Are ATM mutations 7271T->G and IVS10-6T->G really high-risk brest cancer-susceptibility alleles? Cancer Res 2004; 64:840-843.
55. Spring K, Ahangari F, Scott SP et al. Mice heterozygous for mutation in atm, the gene involved in ataxia telangiectasia, have heightened susceptibility to cancer. Nat Genet 2002; 32:185-190.
56. Curry CJ, O'Lague P, Tsai J et al. ATFresno: a phenotype linking ataxia telangiectasia with the nijmegen breakage syndrome. Am J Hum Genet 1989; 45:270-275.
57. Pagani F, Buratti E, Stuani C et al. A new type of mutation causes a splicing defect in ATM. Nat Genet 2002; 30:426-429.
58. Lewandowska MA, Stuani C, Parvizpur A et al. Functional studies on the ATM intronic splicing processing element. Nucl Acids Res 2005; 33:4007-4015.
59. Eng L, Coutinho G, Nahas S et al. Nonclassical splicing mutations in the coding and noncoding regions of the ATM gene: maximum entropy estimates of splice junction strengths. Hum Mutat 2004; 23:67-76.

Chapter 3

Cell Signaling in Ataxia Telangiectasia

Tetsuo Nakajima*

Abstract

Ataxia telangiectasia (AT) is a disease with pleiotropic defects that include hypersensitivity to ionizing radiation, immunodeficiency and increased cancer risk and Ataxia Telangiectasia Mutated (ATM) gene is responsible for AT. Particularly since AT patients are strongly radiosensitive, the relationship between ATM and DNA damage has been studied thoroughly. ATM recognizes the DNA double-strand breaks produced by DNA damaging agents such as ionizing radiation and subsequently controls cell fate by DNA repair, cell cycle regulation and apoptosis. Cell signaling cascades involving ATM are diverse and in terms of radiation-induced apoptosis, numerous participants and their roles have been uncovered in recent years. Here, ATM-related signaling cascades are reviewed, focusing on radiation-induced apoptosis. Relationships with other signaling cascades, such as oxidative stress-related and insulin-related signaling are also discussed.

Introduction

Ataxia telangiectasia (AT) is a disease with pleiotropic defects that include hypersensitivity to ionizing radiation, immunodeficiency, increased cancer risk, cerebeller degeneration, growth retardation and oculocutaneous telangiectasias. The mechanism of AT has been analyzed from the viewpoint of response to DNA damage, as AT cells have chromosomal instabilities similar to xeroderma pigmentosum, Bloom syndrome and Fanconi anemia. Since the identification of ataxia telangiectasia mutated (ATM), the gene responsible for AT, substantial research has been performed.[1-8] As AT is caused by the lack or inactivation of ATM, analysis of ATM function is indispensable in understanding the disease. Studies using ATM-deficient mice and cells from AT patients have clarified the role of ATM.[9,10] ATM is thought to mediate initial recognition of DNA double-strand breaks (DSB) caused by DNA-damaging agents, such as X-rays and to subsequently control cell fate via cell cycle control, DNA repair and apoptosis. The roles of ATM in cell cycle regulation in DNA-damaged cells have been analyzed and its relationships with other factors, such as p53, Chk2 and NBS1 (the Nijmegen breakage syndrome protein), all of which are phosphorylated by ATM,[1-3,7] have been elucidated. The transcriptional activity of NF-κB and p53 after exposure to DNA-damaging agents also appears to be controlled by ATM.[11] However, in this chapter relevant aspects of ATM signalling will be discussed.

Although NF-κB and p53 are associated with ATM, they are also involved in the regulation of apoptosis and constitute complex molecular networks. Furthermore, molecules such as protein kinase C, ceramide and JNK, also involved in apoptosis regulation, participate in ATM signaling. Therefore, this review will focus on the roles of ATM in apoptosis regulation as it is important for understanding ATM-related cell signaling and the cell cycle regulation and DNA repair will not be discussed here.[1-8] In particular the focus will be on the participation of ATM signaling in radiation-induced apoptosis. The relationships between ATM function and oxidative stress-related

*Tetsuo Nakajima—Research Center for Radiation Protection, National Institute of Radiological Sciences, Chiba 263-8555, Japan. Email: otetsu@nirs.go.jp

Molecular Mechanisms of Ataxia Telangiectasia, edited by Shamim I. Ahmad.

and insulin-related signaling cascades, with regard to radiosensitivity and apoptosis regulation will also be discussed.

ATM and Radiation-Induced Apoptosis

AT cells are very sensitive to ionizing radiation such as X-rays and radiosensitivity is generally evaluated by reproductive cell death such as clonogenic cell survival. However, radiation-induced apoptosis is also an index of AT. The radiosensitivity of AT cells has mainly been investigated using clonogenic cell survival, but the function of ATM is very unique from the viewpoint of radiation-induced apoptosis. After radiation-induced DSB and following its recognition by MRN complex (Mre11/Rad50/NBS1), ATM carries out the repair process.[8] ATM usually exists in its dimeric form, which dissociates to an autophosphorylated monomer to work with MRN complex when DNA is damaged. Several signaling molecules for ATM have also been identified and their functions have been investigated in relation to ATM signaling[2,8] (see Fig. 1). The functions of various molecules are described below.

ATM and p53

The link between ATM and p53 has been of particular interest because p53 is phosphorylated by ATM.[1,2,4] Although p53 has already been shown to play other roles in radiation-induced apoptosis,[5] ATM appears to participate in p53 apoptosis cascades and the role of NBS1 signaling in apoptosis regulation is indispensable in the ATM-p53 route.[12] Moreover, induction of transcriptional activity of p53 after exposure of ionizing radiation does not appear to be involved in ATM-related apoptosis regulation, instead Chk2 is a key participant in controlling p53 transcriptional activity-dependent apoptosis regulation. For example, Puma expression in radiation-induced apoptosis in the thymus is not linked to ATM function,[12,13] instead it is dependent on Chk2. Although Chk2 is a substrate of ATM, it functions as part of a different cascade. Indeed, p53-dependent apoptosis after irradiation is regulated by ATM and Chk2 in parallel; however, Chk2 signaling is partly dependent on ATM.[12,14] The relationship between ATM and Chk2 and their roles in apoptosis regulation, thus requires further studies.[15]

ATM, Ceramide and JNK

Ceramide is a lipid mediator and one of the best characterised bioactive sphingolipids. It is produced either via the hydrolysis of sphingomyelin by SMase (sphingomyelinase) or by de novo synthesis via CS (ceramide synthase) and acts as a second messenger in cell signaling system. Ceramide has been found to be a critical participant of apoptosis regulation. Since its role was identified in radiation-induced apoptosis, the network has been investigated.[16] Although the relationship between ATM and ceramide in radiation-induced apoptosis regulation has been demonstrated in lymphoblasts,[17] its mechanisms remain still unclear. Radiation has been shown to induce biphasic increases in intracellular ceramide production linked to apoptosis. The first increase is based on SMase, which is DNA damage-independent and the second increase is due to elevated CS activity through ATM signaling, which apparently comes from DNA repair signaling. Interestingly, the second increase of ceramide is dependent on the first one. SMase regulates apoptosis but it is not related to cell cycle arrest after irradiation. Ceramide production by SMase and CS after irradiation also occurs in other cells such as bovine aortic endothelial cells.[16,18]

Although radiation-induced activation of c-Jun NH_2-terminal kinase (JNK) is likely to be involved in ATM/ceramide cascade, the mechanism of JNK function remains unclear. The regulation may involve the process of sensing of oxidative stress.[19] Radiation and oxidative stress induce JNK activation. In AT cells, oxidative stress induces biphasic activation of JNK and interestingly, the second phase of endogenous c-Jun phosphorylation is more markedly induced in AT cells than in the control cells. Ceramide induces JNK activation and ceramide and JNK appear to be coordinated in controlling apoptosis.[20] In AT cells, radiation-induced JNK activation is inhibited, however, as ATM down-regulation also leads to increased JNK activity,[21] ATM may play a role in the regulation of stress-induced JNK activation via ceramide production and sensing of oxidative stress.

Cross-Talk between PKC and ATM in Radiation-Induced Apoptosis

Protein kinase C (PKC) is a family of serine/threonine kinases, of which there are at least 10 subtypes. Many of the PKC-related signaling molecules for apoptosis have recently been identified;[22,23] ATM is one of these, which regulates PKC-related radiation-induced apoptosis. In mouse thymocytes and thymic lymphomas, radiation-induced apoptosis depends on Atm.[9,24] ATM also participates in PKCδ regulation via c-Abl in response to DNA damage and appears to control the subsequent PKCδ degradation by caspase-3, which results in apoptosis.[25] We have also demonstrated that ATM plays a role in the regulation of radiation-induced apoptosis via PKCδ degradation.[24]

Conversely, ATM inhibits radiation-induced apoptosis in some cases. Down-regulation of ATM protein by treatment with TPA or antisense-ATM oligonucleotides promotes radiation-induced apoptosis in prostate cancer cells,[26] in which ATM normally inhibits ceramide synthase. This inhibition by ATM prevents the cells from entering the apoptotic stage, while treatment with TPA promotes ceramide production. ATM down-regulation appears to mediate this TPA-induced promotion.

TPA is a PKC activator and thus PKC is likely to have a role in mediating radiation-induced apoptosis signals via ATM. Indeed, PKCδ activation results in the down-regulation of ATM protein.[26] ATM down-regulation forces cells to enter an apoptotic stage, but actual apoptosis does not occur. Apoptosis induction requires another apoptosis inducer (such as radiation) after ATM down-regulation. These observations suggest that regulation of ATM-related radiation-induced apoptosis via PKC is tissue- and cell-type-specific. Some of PKC subtypes also have anti-apoptotic effects in the regulation of radiation-induced apoptosis.[23] The mediation of PKC in radiation-induced apoptosis via ATM is thus a challenging topic.

ATM Function and NF-κB Signaling after Irradiation

The link between ATM and NF-κB is unique in the regulation of genotoxic stress-induced apoptosis.[27] NF-κB is a family of transcription factors that exist as homo- or heterodimers assembled from the subunits, p50, p52, p65, c-Rel and RelB. The heterodimer p50/p65 is most commonly found. After DNA damage, activation of NF-κB occurs which is linked to the complex[28] consisting of PIDD (p53-induced protein with a death domain), RIP1 (receptor-interacting protein 1) and NEMO (NF-κB essential modulator). RIP1 recruits NEMO to PIDD and the complex accumulates in the nucleus. NEMO sumoylation is induced after treatment with DNA-damaging agents. This sumoylation is enhanced when NEMO is in the complex. Sumoyled NEMO is then phosphorylated by ATM, after which NEMO is desumoyled and ubiquitinated and the heterodimers of NEMO and ATM are translocated to the cytosol to join with IKKα and β resulting in activation of IKK. Although many processes in this cascade remain unknown, the association of NEMO and ATM with IKK induces transcriptional activity of p50/p65 leading to cell survival.

In the NF-κB cascade, ATM interacting with NEMO is translocated to the cytoplasm, where it regulates the IKK cascade.[29] However, this cascade occurs at 60 to 90 minutes after stress begins and in comparison with the time of ATM recognition of DSB (within 10 seconds), it appears to be a later step in the regulatory stages of apoptosis.

Recent Research into ATM Function, Radiation-Induced Apoptosis and Radiosensitivity

Alteration of radiosensitivity by mutation of ATM cannot be explained by control of apoptosis alone. Many novel molecules have been detected by the research in DNA repair-linked pathways and ATM-induced carcinogenesis. After DNA damage, ATM initially delays the cell cycle, which allows time for DNA repair and determining cell fate. ATM-deficient cells cannot recognize DSB and DNA state collapses resulting in cell death.

ATM, Clonogenic Survival and Radiosensitivity

Clonogenic survival in radiosensitive cell culture has routinely been used as the main index for AT. Although the factors participating in the signaling of ATM have been identified in recent

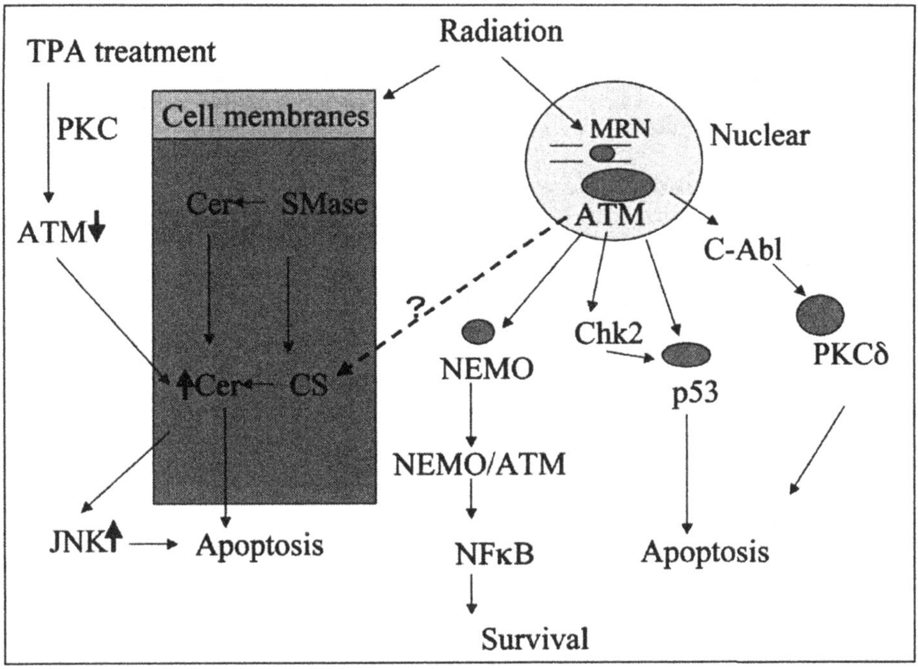

Figure 1. ATM and radiation-induced apoptosis regulation. This figure summarizes ATM-related signaling in radiation-induced apoptosis describing in the text. Cer: ceramide, SMase: sphingomyelinase, CS: ceramide synthase, TPA: 12-O-tetradecanoylphorbol-13-acetate, NF-κB : nuclear factor-κB, Chk2: checkpoint kinase 2, NEMO:NF-κB essential modulator, MRN: Mre11/Rad50/NBS1 complex, JNK: c-Jun NH2-terminal kinase.

years, the molecules involved in radiosensitivity, in terms of clonogenic survival, are yet to be identified. For example, thousand and one amino acid (TAO) kinases are MAP kinase kinase kinases (MAP3Ks) related to ATM signaling. TAO kinases appear to be substrates of ATM and involved in regulation of G2/M arrest. In cells in which TAO kinases have been knocked down, inhibition of DNA damage-induced G2/M checkpoint is displayed. Furthermore, cell's sensitivity to radiation, as assessed by clonogenic survival, increases in these cells.[30] This increased radiosensitivity may be due to disruption of DNA damage-induced cell checkpoint. However, participation of TAO kinases in the regulation of radiation-induced apoptosis remains unknown. Evaluation of radiosensitivity by clonogenic survival such as colony forming assays must be carefully distinguished from that by apoptosis.

ATM, Apoptosis and Radiosensitivity

In ATM-related radiosensitivity, apoptosis is currently the major focus. BAAT1 is another ATM-related molecule and in BAAT1-deficient cells, apoptosis occurs even without DNA damage.[31] Although the mechanisms of stress-induced apoptosis regulation remain unclear, NBS1 may be involved in the ATM-related determinants in inducing radiation-induced apoptosis. NBS1 is an element in the MRN complex, which initially recognizes DSB. It was recently reported that the carboxy terminus (C-terminal domain) of NBS1 is important for radiation-induced apoptosis control.[12,15] Loss of the C-terminal domain of NBS1 inhibits phosphorylation of SMC1 (S-phase regulating protein), which is phosphorylated in an ATM-dependent manner after irradiation. This leads to dysfunction of cell cycle regulation after irradiation and apoptosis is inhibited, presumably because phosphorylation of BID, which is also carried out by ATM, is inhibited in NBS-1 C-terminal domain-deficient cells. However, in these cells, autophosphorylation of ATM serine 1981 is normal,[12] and phosphorylation of NBS1 by ATM is also intact.[15] When the MRN complex involving NBS1 triggers apoptosis induction, it is still likely to require cooperation with ATM;[12] however, the mechanisms are unclear. As ATM movement to DSB occurs normally in NBS1-C-terminal domain-deficient cells,[15] it appears that the NBS1-C-terminal domain regulates ATM access to its substrates for determination of cell fate.

ATM Signaling and Oxidative Stress

In AT patients, oxidative stress is thought to lead to carcinogenesis, neurodegeneration and premature aging.[32] It has been reported that oxidative stress is elevated (e.g., a rise in the intracellular levels of reactive oxygen species) in Atm-deficient mice and cells of AT patient origin.[32] In Atm-deficient mice, there is a high incidence of thymic lymphomas; however, the neurodegenerative changes (e.g., cerebellar degeneration), that are characteristic of AT, are different in deficient mice; Atm-deficient mice have no histological brain abnormalities, but they have neurological behavioral abnormalities.[33] As cells in the thymus and brain of Atm-deficient mice are radioresistant in terms of radiation-induced apoptosis, oxidative stress may induce AT phenotypes via defects of stress-induced apoptosis regulation.[9,34,35] However, oxidative stress might not necessarily be related to the development of thymic lymphomas or neuronal degeneration.[36-38]

Oxidative Stress, Apoptosis and Carcinogenesis

The link between carcinogenesis and apoptosis has also been investigated in NBS1 C-terminal-deficient mice. Atm-dependent apoptosis regulation is dependent on the NBS1 C-terminus, as discussed above. Thymic lymphomas do not develop in NBS1 C-terminal-deficient mice in contrast to Atm-deficient mice, while cells in the thymus of both types of mutant mice strains are resistant to radiation-induced apoptosis. This confirms that apoptosis is not involved in development of thymic lymphomas.[12]

There are several reports regarding the link between ATM and oxidative stress. For example, SAT (SOD and AT) mice, which are deficient for Atm, that overexpress the human SOD1 (Cu/Zn-superoxide dismutase) protein, have been particularly useful for investigating the link between oxidative stress and ATM.[38] In these mice, SOD1 is overexpressed and intracellular oxidative stress is accelerated. This results in increased susceptibility to radiation and abnormal T-cell differentiation. However, the incidence of thymic lymphomas is the same as in $Atm^{-/-}$ mice.[38] Therefore, oxidative stress is not likely to be linked to carcinogenesis; however, antioxidants delay the onset of thymic lymphoma or even suppress lymphomas in $Atm^{-/-}$ mice,[35,36,39,40] and the mutational pattern of AT is similar to the oxidative stress-induced mutation pattern.[41] These reports imply a link between oxidative stress and carcinogenesis

Spontaneous, Oxidative Stress-Induced and Radiation-Induced Apoptosis

In addition to carcinogenesis, spontaneous apoptosis in the thymic cells is promoted in $Atm^{-/-}$ mice;[10,42] although radiation-induced apoptosis in the cells from these mice is inhibited.[9] Spontaneous apoptosis is promoted in certain cell types of AT patients origin, while oxidative stress-induced apoptosis is inhibited.[43] In contrast, $Atm^{-/-}$ cells with DT40 are sensitive both to radiation and oxidative stress.[44] The link between ATM function and apoptosis regulation remains to be investigated from the viewpoint of oxidative stress. Cell sensitivity to radiation and oxidative stress, in terms of apoptosis, is cell specific as described above and the mechanisms will have to be thoroughly investigated.[45]

ATM Distribution, Oxidative Stress and AT Phenotype

The link between ATM and oxidative stress in cerebellar tissue is also interesting. The intracellular distribution and function of ATM are controversial,[46-48] as it has been reported that ATM is also present in the cytoplasm of cerebellar tissues.[49,50] It is mainly seen in the nucleus, particularly in proliferating cells.[51] However, small amount of ATM also appears in the cytoplasm; for example in 293T cells, about 30% of ATM can be identified in the cytoplasm.[52] As discussed above, in NF-kB signaling, ATM translocation to the cytosol is important in the regulation of apoptosis.

The distribution of ATM may thus be linked to its distinct functions; for example, ATM interacts with β-adaptin, which participates in vesicle transport.[53]

ATM is also present around peroxisomes, which regulate intracellular redox-state via catalase.[54] In AT, catalase activity is decreased and lipid peroxidation increased.[54] These findings suggest that ATM function is related to oxidative stress regulation. Moreover, these defects in oxidative stress regulation may explain why Atm-deficient mice exhibit neurological behavioral abnormalities, despite having a morphologically normal nervous system.[50]

ATM function in cerebellar tissue is probably also linked to DNA damage recognition, which occurs in the nucleus, not in the cytoplasm.[46] On the other hand, in $Atm^{-/-}$ mouse brain, although apoptosis does not occur,[35] DNA damage is increased and antioxidants reduce the development of neurological abnormalities.[36] Taken together, these observations suggest that, in Atm-deficient mice, elevated oxidative stress damages DNA and subsequently induces nerve damage, but not via apoptosis.

ATM and Insulin Signaling

If the symptoms of AT develop in childhood, growth delay and glucose intolerance are observed in patients. However, the mechanisms remain unknown. Some AT patients suffer from insulin resistance and dysfunctional glycemic regulation. Decreased ATM expression increases insulin resistance and atherosclerosis rates. This has been demonstrated by experiments on $apoE^{-/-}$ mice, a well established animal model for atherosclerosis.[21] As macrophages initiate early atherosclerotic lesions, in the study, $ATM^{+/+}$ $apoE^{-/-}$ mice were transplanted with bone marrow from $ATM^{+/+}$ $apoE^{-/-}$ or $ATM^{-/-}$ $apoE^{-/-}$ mice. Results showed that the atherosclerosis lesions were 80% larger in animals transplanted with bone marrow from ATM null marrow. Furthermore, insulin resistance in $Atm^{+/-}$ mice was increased in comparison with $Atm^{+/+}$ mice. It was also demonstrated that chloroquine, an antimalarial agent, activates ATM and decreases glucose levels.[21,55] These effects of chloroquine were most likely due to inhibition of insulin degradation or increased insulin secretion.[21,56,57] Thus, further studies of ATM signaling may explain the mechanisms of atherosclerosis and insulin resistance in metabolic syndrome.

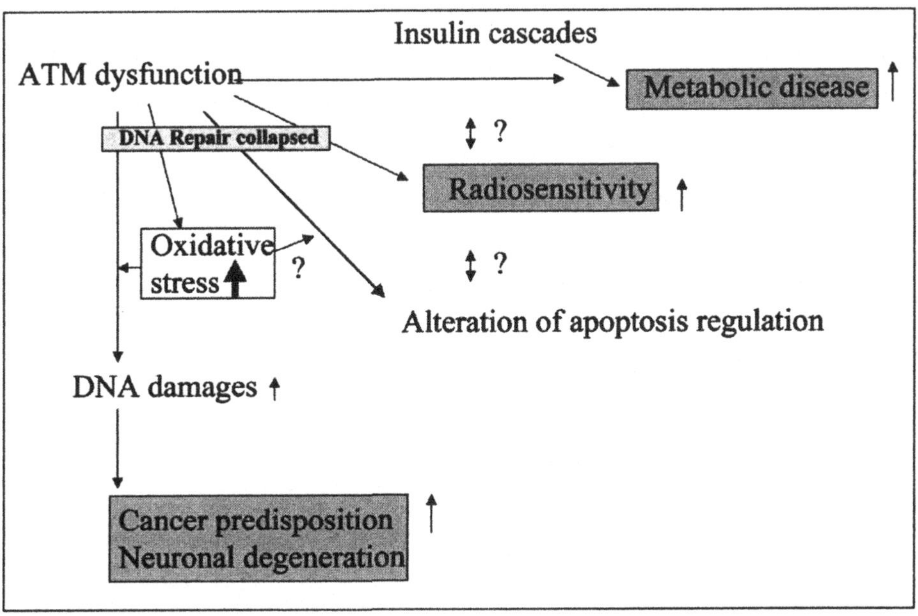

Figure 2. ATM function and AT phenotypes.

Other studies to determine the link between ATM and insulin have shown that phosphorylation of translation factors activated by insulin is influenced by ATM and that ATM kinase activity is enhanced by insulin treatment.[52,58] It has also been reported that activation of PKB/Akt is mediated by ATM signaling after treatment with insulin or radiation.[59]

In addition to insulin, insulin-like growth factor-1 (IGF-1)-related signaling may be linked to ATM signaling. The functions of insulin and IGF-1 appear to have some overlaps.[60-62] ATM is probably associated with insulin-like growth factor-1 receptor (IGF-1R) in breast cancer radioresistance.[63] Low levels of IGF-1R expression augments radiosensitivity in AT cells. Complementation of AT cells with the ATM cDNA increases IGF-1R expression, resulting in increased radioresistance.[64] Further, studies of insulin and IGF-1 signaling may provide new insight into ATM function.

Perspectives in ATM Signaling Research

In this chapter, ATM signaling has been reviewed with the main focus on radiation-induced apoptosis. ATM appears to control radiation-induced apoptosis in many distinct stages and an important finding on ATM function indicate that its regulation is a well-organized biological network. Therefore, mutation leading to dysfunction of ATM regulation leads to AT. A link between ATM signaling and AT phenotypes are summarized in Figure 2. ATM may play a role in cross-talk between insulin-related signaling and radiation-induced damage signaling. Moreover, possibility that ATM function in relation to oxidative stress and insulin signaling may play a role at extranuclear sites suggests that topological analysis is needed to fully elucidate the mechanisms leading to AT. Recent evidence suggesting that extracellular mediators such as TGF-β are required for ATM recognition of DSB[65] and hence ATM must interact with other network participants to function. Although network analysis of ATM and related molecules is still needed, cross-talk and interactive associations with other biomolecules are indispensable for understanding the ATM signaling network, and although elucidating the whole range of ATM functions will require substantial research, the diversity of ATM functions will continue to attract interest for some years to come.

References

1. Shiloh Y. ATM and ATR: networking cellular responses to DNA damage. Curr Opin Gent Dev 2001; 11:71-77.
2. Shiloh Y, Kastan MB. ATM: genome stability, neuronal development and cancer cross paths. Adv Cancer Res 2001; 83:209-254.
3. Abraham RT. Cell cycle checkpoint signaling through the ATM and ATR kinases. Genes Dev 2001; 15:2177-2196.
4. Lavin MF, Shiloh Y. The genetic defect in ataxia telangiectasia. Annu Rev Immunol 1997; 15:177-202.
5. Meyn MS. Ataxia telangiectasia, cancer and the pathobiology of the ATM gene. Clin Genet 1999; 55:289-304.
6. Lavin MF, Birrell G, Chen P et al. ATM signaling and genomic stability in response to DNA damage. Mutat Res 2005; 569:123-132.
7. Lavin MF, Khanna. KK. Review: ATM : the protein encoded by the gene mutated in the radiosensitive syndrome ataxia telangiectasia. Int J Radiat Biol 1999; 75:1201-1214.
8. Kurz EU, Lees-Miller SP. DNA damage-induced activation of ATM and ATM-dependent signaling pathways. DNA Repair 2004; 3:889-900.
9. Xu Y, Baltimore D. Dual roles of ATM in the cellular response to radiation and in cell growth control. Genes Dev 1996; 10:2401-2410.
10. Elson A, Wang Y, Daugherty CJ et al. Pleiotropic defects in ataxia telangiectasia protein-deficient mice. Proc Natl Acad Sci USA 1996; 93:13084-13089.
11. Rashi-Elkeles S, Elkon R, Weizman N et al. Parallel induction of ATM-dependent pro and antiapoptotic signals in response to ionizing radiation in murine lymphoid tissue. Oncogene 2006; 25:1584-1592.
12. Stracker TH, Morales M, Couto SS et al. The carboxy terminus of NBS1 is required for induction of apoptosis by the MRE11 complex. Nature 2007; 447:218-221.
13. Villunger A, Michalak EM, Coultas L et al. p53- and drug-induced apoptotic responses mediated by BH3-only proteins Puma and Noxa. Science 2003; 302:1036-1038.

14. Hirao A, Cheung A, Duncan G et al. Chk2 is a tumor suppressor that regulates apoptosis in both an Ataxia Telangiectasia Mutated (ATM)-dependent and an ATM-independent manner. Mol Cell Biol 2002; 22:6521-6532.
15. Difilippantonio S, Celeste A, Kruhlak M et al. Distinct domains in Nbs1 regulate irradiation-induced checkpoints and apoptosis. J Exp Med 2007; 204:1003-1011.
16. Haimovitz-Friedman A, Kan CC, Ehleiter D et al. Ionizing radiation acts on cellular membranes to generate ceramide and initiate apoptosis. J Exp Med 1994; 180:525-535.
17. Vit JP, Rosselli F. Role of the ceramide-signaling pathways in ionizing radiation-induced apoptosis. Oncogene 2003; 22:8645-8652.
18. Liao WC, Haimovitz-Friedman A, Persaud RS et al. Ataxia telangiectasia-mutated gene product inhibits DNA damage-induced apoptosis via ceramide synthase. J Biol Chem 1999; 274:17908-17917.
19. Lee SA, Dritschilo A, Jung M. Role of ATM in oxidative stress-mediated c-Jun phosphorylation in response to ionizing radiation and $CdCl_2$. J Biol Chem 2001; 276:11783-11790.
20. Verheij M, Bose R, Lin XH et al. Requirement for ceramide-initiated SAPK/JNK signaling in stress-induced apoptosis. Nature 1996; 380:75-79.
21. Schneider JG, Finck BN, Ren J et al. ATM-dependent suppression of stress signaling reduces vascular disease in metabolic syndrome. Cell Metabolism 2006; 4:377-389.
22. Musashi M, Ota S, Shiroshita N. The role of protein kinase C isoforms in cell proliferation and apoptosis. Int J Hematol 2000; 72:12-19.
23. Nakajima T. Signaling cascades in radiation-induced apoptosis: Roles of protein kinase C in the apoptosis regulation. Med Sci Monit 2006; 12:220-224.
24. Nakajima T, Yukawa O, Tsuji H et al. Regulation of radiation-induced protein kinase Cδ activation in radiation-induced apoptosis differs between radiosensitive and radioresistant mouse thymic lymphoma cell lines. Mutat Res 2006; 595:29-36.
25. Yoshida K, Wang HG, Miki Y et al. Protein kinase Cδ is responsible for constitutive and DNA damage-induced phosphorylation of Rad9. EMBO J 2003; 22:1431-1441.
26. Truman JP, Gueven N, Lavin M et al. Down-regulation of ATM protein sensitizes human prostate cancer cells to radiation-induced apoptosis. J Biol Chem 2005; 280:23262-23272.
27. Habraken Y, Piette J. NF-κB activation by double-strand breaks. Biochem Pharmacol 2006; 72:1132-1141.
28. Lin Y, Ma W, Benchimol S. Pidd, a new death-domain-containing protein, is induced by p53 and promotes apoptosis. Nature Genet 2000; 26:124-127.
29. Wu ZH, Shi Y, Tibbetts RS et al. Molecular linkage between the kinase ATM and NF-κB signaling in response to genotoxic stimuli. Science 2006; 311:1141-1146.
30. Raman M, Earnest S, Zhang K et al. TAO kinases mediate activation of p38 in response to DNA damage. EMBO J 2007; 26:2005-2014.
31. Aglipay JA, Martin SA, Tawara H et al. ATM activation by ionizing radiation requires BRCA1-associated BAAT1. J Biol Chem 2006; 281:9710-9718.
32. Barzilai A, Rotman G, Shiloh Y. ATM deficiency and oxidative stress: a new dimension of defective response to DNA damage. DNA Repair 2002; 1:3-25.
33. Barlow C, Hirotsune S, Paylor R et al. Atm-deficient mice: A paradigm of ataxia telangiectasia. Cell 1996; 86:159-171.
34. Herzog KH, Chong MJ, Kapsetaki M et al. Requirement for Atm in ionizing radiation-induced cell death in the developing central nervous system. Science 1998; 280:1089-1091.
35. Stern N, Hochman A, Zemach N et al. Accumulation of DNA damage and reduced levels of nicotine adenine dinucleotide in the brains of Atm-deficient mice. J Biol Chem 2002; 277:602-608.
36. Gueven N, Luff J, Peng C et al. Dramatic extension of tumor latency and correction of neurobehavioral phenotype in Atm-mutant mice with a nitroxide antioxidant. Free Radic Biol Med 2006; 41:992-1000.
37. Erker L, Schubert R, Elchuri S et al. Effect of the reduction of superoxide dismutase 1 and 2 or treatment with α-tocopherol on tumorigenesis in Atm-deficient mice. Free Radic Biol Med 2006; 41:590-600.
38. Peter Y, Rotman G, Lotem J et al. Elevated Cu/Zn-SOD exacerbates radiation sensitivity and hematopoietic abnormalities of Atm-deficient mice. EMBO J 2001; 20:1538-1546.
39. Reliene R, Schiestl RH. Antioxidant N-acetyl cysteine reduces incidence and multiplicity of lymphoma in Atm deficient mice. DNA Repair 2006; 5:852-859.
40. Reliene R, Schiestl RH. Antioxidants suppress lymphoma and increase longevity in Atm-deficient mice. J Nutr 2007; 137:229S-232S.
41. Gage BM, Alroy D, Shin CY et al. Spontaneously immortalized cell lines obtained from adult Atm null mice retain sensitivity to ionizing radiation and exhibit a mutational pattern suggestive of oxidative stress. Oncogene 2001; 20:4291-4297.

42. Yan M, Kuang X, Qiang W et al. Prevention of thymic lymphoma development in Atm−/− mice by dexamethasone. Cancer Res 2002; 62:5153-5157.
43. Schubert R, Reichenbach J, Royer N et al. Spontaneous and oxidative stress-induced programmed cell death in lymphocytes from patients with ataxia telangiectasia (AT). Clin Exp Immunol 2000; 119:140-147.
44. Takao N, Li Y, Yamamoto K. Protective roles for ATM in cellular response to oxidative stress. FEBS Lett 2000; 472:133-136.
45. Duchaud E, Ridet A, Stoppa-Lyonnet D et al. Deregulated apoptosis in ataxia telangiectasia:Association with clinical stigmata and radiosensitivity. Cancer Res 1996; 56:1400-1404.
46. Dar I, Biton S, Shiloh Y et al. Analysis of the ataxia telangiectasia mutated-mediated DNA damage response in murine cerebellar neurons. J Neurosci 2006; 26:7767-7774.
47. Biton S, Dar I, Mittelman L et al. Nuclear ataxia telangiectasia mutated (ATM) mediates the cellular response to DNA double strand breaks in human neuron-like cells. J Biol Chem 2006; 281:17482-17491.
48. Biton S, Gropp M, Itsykson P et al. ATM-mediated response to DNA double strand breaks in human neurons derived from stem cells. DNA Repair 2007; 6:128-134.
49. Oka A, Takashima S. Expression of the ataxia telangiectasia gene (ATM) product in human cerebellar neurons during development. Neuroscience Lett 1998; 252:195-198.
50. Barlow C, Ribaut-Barassin C, Zwingman TA et al. ATM is a cytoplasmic protein in mouse brain required to prevent lysosomal accumulation. Proc Natl Acad Sci USA 2000; 97:871-876.
51. Brown KD, Ziv Y, Sadanandan SN et al. The ataxia telangiectasia gene product, a constitutively expressed nuclear protein that is not up-regulated following genome damage. Proc Natl Acad Sci USA 1997; 94:1840-1845.
52. Yang DQ, Kastan MB. Participation of ATM in insulin signalling through phosphorylation of eIF-4E-binding protein 1. Nature Cell Biol 2000; 2:893-898.
53. Lim DS, Kirsch DG, Canman CE et al. ATM binds to β-adaptin in cytoplasmic vesicles. Proc Natl Acad Sci USA 1998; 95:10146-10151.
54. Watters D, Kedar P, Spring K et al. Localization of a portion of extranuclear ATM to peroxisomes. J Biol Chem 1999; 274:34277-34282.
55. Shoelson SE. Banking on ATM as a new target in metabolic syndrome. Cell Metabolism 2006; 4:337-338.
56. Asamoah KA, Robb DA, Furman BL. Chronic chloroquine treatment enhances insulin release in rats. Diabetes Res Clin Pract 1990; 9:273-278.
57. Blazar BR, Whitley CB, Kitabchi AE et al. In vivo chloroquine-induced inhibition of insulin degradation in a diabetic patient with severe insulin resistance. Diabetes 1984; 33:1133-1137.
58. Lavin MF. An unlikely player joins the ATM signalling network. Nature Cell Biol 2000; 2:E215-217.
59. Viniegra JG, Martínez N, Modirassari P et al. Full activation of PKB/Akt in response to insulin or ionizing radiation is mediated through ATM. J Biol Chem 2005; 280:4029-4036.
60. Lammers R, Gray A, Schlessinger J et al. Differential signalling potential of insulin- and IGF-1-receptor cytoplasmic domains. EMBO J 1989; 8:1369-1375.
61. Lai A, Sarcevic B, Prall OWJ et al. Insulin/Insulin-like growth factor-1 and estrogen cooperate to stimulate cyclin E-Cdk2 activation and cell cycle progression in MCF-7 breast cancer cells through differential regulation of cyclin E and $p21^{WAF1/Cip1}$. J Biol Chem 2001; 276:25823-25833.
62. Pandini G, Vigneri R, Costantino A et al. Insulin and insulin-like growth factor-1 (IGF-1) receptor overexpression in breast cancers leads to insulin/IGF-1 hybrid receptor overexpression: Evidence for a second mechanism of IGF-1 Signaling. Clin Cancer Res 1999; 5:1935-1944.
63. Jameel JKA, Rao VSR, Cawkwell L et al. Radioresistance in carcinoma of the breast. The Breast 2004; 13:452-460.
64. Peretz S, Jensen R, Baserga R et al. ATM-dependent expression of the insulin-like growth factor-1 receptor in a pathway regulating radiation response. Proc Natl Acad Sci USA 2001; 98:1676-1681.
65. Kirshner J, Jobling MF, Pajares MJ et al. Inhibition of transforming growth factor-β1 signaling attenuates ataxia telangiectasia mutated activity in response to genotoxic stress. Cancer Res 2006; 66:10861-10869.

CHAPTER 4

DNA Damage and Repair in Ataxia Telangiectasia

Melissa M. Adams and Phillip B. Carpenter*

Abstract

Defects in the gene Ataxia Telangiectasia Mutated (ATM) are responsible for the development of ataxia telangiectasia (AT), an incurable cancer-prone disease that is accompanied by a pleiotropic set of conditions including, amongst others, neurological and immunological deficiencies. ATM is a large protein kinase that controls numerous cellular processes, particularly those involved with maintaining chromosome stability. In recent years, a wealth of data have shown that ATM regulates multiple events throughout the cell cycle, including replication origin firing, telomere dysfunction and programmed events at antigen receptor loci in T-and B-cells of lymphoid origin. As all of these processes utilize DNA-double stranded break (DSB) intermediates, ATM suppresses the generation of DSB-inducing translocation events through regulating multiple pathways. Here, we discuss the activation and role of ATM kinase and its closely associated factors in maintaining chromosomal stability in various DSB signaling pathways. Notably, we highlight some recent advances suggesting that ATM suppresses chromosomal translocations in lymphoid cells through its role in maintaining free DNA ends during DNA end joining, a process that links DSB repair, chromatin function and, the development of lymphoma.

Introduction

Ataxia telangiectasia (AT) is an autosomal recessive and incurable disease afflicting all races. AT is characterized by multiple, well-documented phenotypes including ocular telangiectasias ("spider veins"), awkward gait, slurred speech, progressive neurodegeneration, metabolic disorders, immune deficiencies and genomic instability. AT has an incidence of ~1:40,000-100,000 and the carrier frequency is between 1:100-200. AT patients and cells derived from them are sensitive to ionizing radiation (IR) and defective in multiple cell cycle checkpoints. Originally localized to chromosome 11q22-23 by linkage analysis, the gene defective in AT, mutated in ataxia telangiectasia (ATM), was identified through a positional cloning strategy and found to encode a large protein of 370 kD that belongs to the phosphoinositide 3 kinase (PIK) superfamily.[1] However, although ATM possesses a PIK domain, it phosphorylates proteins rather than lipids. Additionally, two other PIK family members, the DNA-dependent protein kinase (DNA-PK) and ATR (ataxia telangiectasia and Rad3-related) are intimately related to ATM function and the DNA damage response. Since its initial, seminal identification in 1995 by Shiloh and colleagues,[1] a large body of exciting research, employing genetic, biochemical, molecular and structural techniques in various model systems, has demonstrated that ATM acts as a master regulator of several pathways, particularly those that maintain chromosomal stability in response to DNA-double strand breaks

*Corresponding Author: Phillip B. Carpenter—Department of Biochemistry and Molecular Biology, University of Texas Health Science Center, Houston, TX 77030, USA.
Email: phillip.b.carpenter@uth.tmc.edu

Molecular Mechanisms of Ataxia Telangiectasia, edited by Shamim I. Ahmad.

(DSBs). Thus, the ability of ATM to regulate various biological pathways helps to explain the pleiotropic phenotypes displayed by AT patients. Because ATM influences many cellular systems, research into the biological function of ATM has attracted widespread attention from investigators in traditionally diverse fields such as neurobiology, immunology, metabolism and cell cycle control. Therefore, a greater understanding of the mechanism of ATM function will not only help in the designing of the therapeutics to treat the presently incurable AT disease, but will also provide vast knowledge in DNA damage signaling, chromatin function, immune and neuronal development and other processes such as metabolic syndrome(s). Also important questions, such as how all of the pathologies associated with AT are linked to the underlying genetic defect in ATM, may be answered.

ATM and Its Central Role in DSB Repair

AT is one of several autosomal recessive, chromosomal instability syndromes characterized by a strong predisposition to generate cancer. Such disorders include Fanconi anemia (FA), Bloom's syndrome (BS), Nijmegen breakage syndrome (NBS) and ataxia telangiectasia-like disorder (ATLD). NBS and ATLD are caused by hypomorphic mutations in the *NBS1* and *MRE11* genes respectively and these gene products, as discussed below, are intimately related to the biology of ATM kinase. Although distinct, these diseases have overlapping clinical features and as predicted, the underlying genetic defects encode factors with common biochemical functions. For example, like AT, those patients with NBS display extreme sensitivity to IR and possess a strong tendency to develop lymphoid tumors. However, NBS patients, unlike those with AT, do not display neurodegeneration although they often display signs of microencephaly. As the molecular pathologies underlying the various chromosomal instability syndromes have been intensively studied and characterized in both human and mouse models, it has become increasingly clear that the biochemical pathways perturbed in such diseased states are highly convergent in the DNA damage response, particularly the cellular response to DSBs.

DSBs are repaired by either of the two major pathways: (i) homologous recombination (HR) that utilizes a sister chromatid template, generated during S/G_2 and (ii) classical, non homologous end joining (NHEJ), an error prone process that joins broken DNA ends, primarily in G_1. Because HR uses an identical sister chromatid (S/G_2) to provide any lost genetic information through DNA damage, it is considered to be essentially error free. In contrast, NHEJ joins "free" ends, is error prone and is mediated by a ubiquitous set of core proteins that ligate free ends.[2-4] The error prone nature of NHEJ in T- and B-cells significantly contributes to the generation of antibody diversity.[2-4] To ensure the fidelity of chromosomal transmission from generation to generation, cells coordinate DSB repair with chromatin structure and function, in addition to other processes such as transcription and apoptosis. DSBs are formed during normal cell metabolism at stalled replication forks or during T- and B-cell development. DSBs are also generated in response to external agents such as IR and other radiomimetic chemicals (e.g., adriamycin and etoposide). Moreover, dysfunctional telomeres, as produced through loss of the end-protecting shelterin function, or through telomerase-deficient attrition, also activate DSB repair pathways. In response to DSBs, phosphatidylinositol-like kinase (PIK) kinases such as ATM, ATR (ATM and Rad3-related) and DNA-PK (DNA dependent protein kinase) are activated and phosphorylate various substrates primarily on S/TQ motifs. This promotes DNA repair and related processes by modulating chromatin structure and function. ATM maintains genomic stability through the regulation of multiple types of DSB repair response pathways (Fig. 1). ATM and DNA-PK are primarily activated in response to DSBs, but ATR is activated in response to single-stranded DNA (ssDNA). Because ssDNA is often an intermediate of DSB repair processing, typically created through the de-coupling of helicase-polymerase function at stalled replication forks, ATR plays a major role in S-phase and may also be activated by DSBs. In fact, in *Xenopus* cell-free extracts ATM mediates the activation of ATR in response to DSBs, but not to replication stress and this involves the phosphorylation of the ATR-activating protein TopBP1 at S1131.[5] Thus, ATM contributes to ATR activation through two collaborative mechanisms: (i) recession of DNA ends generating

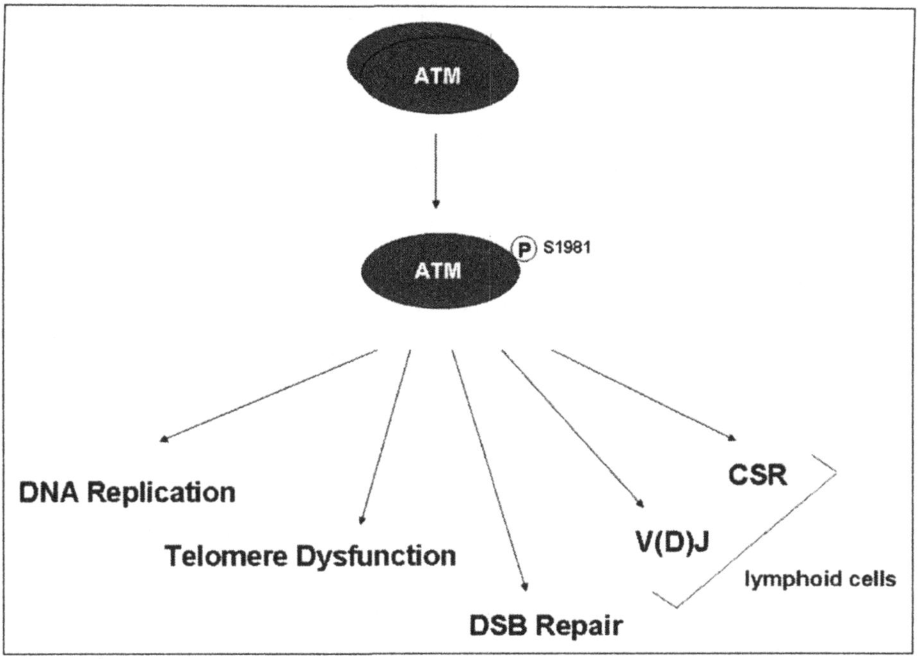

Figure 1. ATM maintains genomic stability through the regulation of multiple types of DSB repair response pathways. Activation of ATM results in its conversion from an inactive dimer to an active monomer. This appears to be accompanied by autophosphorylation at S1981.

ssDNA and (ii) phosphorylation of TopBP1. Additionally, as ssDNA is a natural component of telomeres, ATR also plays an important role in the response to dysfunctional telomeres.[6,7] Unlike ATM, ATR null-mutant mice display embryonic lethality and humans with Seckel syndrome have been found to possess ATR hypomorphism.[8] Importantly, ATM and ATR control the activation of the downstream effector kinases Chk1 and Chk2. In turn, these kinases have been shown to regulate cyclin dependent kinases (Cdks) through the control of M-phase promoting Cdc25 phosphatase activity.

The availability of cells derived from AT patients, especially when considered in combination with AT murine models (see below), has resulted in a cornucopia of knowledge in multiple areas including DNA repair and the cell cycle. Early studies highlighted the role of AT and its ability to either sense and/or respond to DNA damage. Later, numerous subsequent reports have demonstrated that ATM performs functions in both the response and repair of DNA damage. Recently, a large-scale proteomic analysis of proteins phosphorylated at S/TQ sites during the DNA damage response identified more than 700 proteins targeted by ATM and revealed the existence of additional protein modules and networks previously unsuspected of functioning in the DNA damage response.[9] This includes the AKT-insulin pathway as well as factors known to function in chromatin modification, transcription and RNA metabolism and reveals that the DNA damage response profoundly influences numerous cellular processes, many of which were not believed to be mechanistically linked. Thus, the cellular responses to DNA damage (and hence pathways controlled by ATM) are extensive.

Chromosomal Instability in Ataxia Telangiectasia Cells: Implications for Cell Cycle Control

In culture, AT cells have high levels of chromosomal aberrations and are hypersensitive to IR and other radiomimetic agents that induce DSBs. Thus they fail to enforce cell cycle checkpoint pathways after treatment with these agents. This explains the adverse response of AT patients to radiotherapy. Indeed, AT cells show checkpoint defects in the G_1/S and G_2/M transitions. During S-Phase, AT cells possess increased levels of DNA replication intermediates. Thus, prior to the discovery of ATM as a master regulator of the DNA damage response (DDR), it was clear that the gene product, defective in AT, would be involved in maintaining genomic stability. Defects in ATM function are well known to influence the G_1/S transition. Importantly, Kastan et al demonstrated that the DNA damage-dependent phosphorylation of p53 at serine 15 (S15), a modification that contributes to its stabilization against Mdm2-mediated degradation, is delayed in AT cells.[10] Additionally, ATM activates Chk2 and this effector kinase phosphorylates p53 at S20, an event that also contributes to p53 stabilization.[11,12] Yet, ATM targets Mdm2, indicating that the protein enforces the G_1/S checkpoint by controlling a regulatory loop consisting of p53, Chk2 and Mdm2. However, as both Chk2–/– and p53–/– mice exhibit fundamental differences from $ATM^{-/-}$ animals,[12-15] especially with respect to V(D)J recombination in T- and B-cells (see below), ATM appears to control other aspects of the G_1/S checkpoint control outside of the ATM-Chk2-p53 regulatory loop and as discussed below, this is important in NHEJ at antigen receptor loci in B- and T-cells.

ATM, the Intra S-Phase Checkpoint and the Regulation of Origin Firing

Painter and Young[16] showed that after treatment with IR, AT cells continued to synthesize DNA during S-phase. Such a radioresistant DNA synthesis (RDS) phenotype formed the basis for the discovery of the intra S-phase checkpoint. Normally, cells experiencing DNA damage in S-phase inhibit replication origin firing while the DNA damage is repaired. This is often referred to as the intra S-phase checkpoint (Fig. 2). The role of ATM in this process has been shown to influence parallel pathways.[17,18] For example, ATM phosphorylates Nbs1 at S343 in response to IR and this event has been shown to function in maintaining the intra S-phase checkpoint.[19] Along these lines, an ATM-Nbs1-Brca1 pathway has been shown to function in maintaining the S-phase checkpoint through the phosphorylation of the cohesion Smc1 at S957 and S966.[20,21] One reason this is interesting because, as described below, Nbs1 has also been shown to be important in ATM activation, indicating that Nbs1 operates both "upstream" and "downstream" in the ATM pathway. The role of Nbs1 in ATM activation is discussed below. Secondly, through an ATM-Chk2-Cdc25a signaling pathway, ATM influences the activation of origins of replication through negative regulation of Cdk2. This is discussed below.

ATM and Origin Firing in S-Phase

The mechanisms by which ATM (and ATR) regulates origin firing are beginning to be clarified and reveal some important differences between origins of DNA replication ("replicators") in various organisms. Classical studies in prokaryotes and yeast have shown that defined, cis-acting elements, oriC and ARS, respectively and trans-acting factors, dnaA, ORC, respectively, conform to the replicator hypothesis proposed by Jacob and colleagues. Here, trans-acting factors bind cis-acting regulatory elements and position the replication machinery for DNA synthesis. However, more recent studies in *Xenopus* and mammalian systems have suggested that origins in these species are dynamically selected, perhaps by an "origin interference" or "self-regulating" model,[22] suggesting that bona fide origins are not required for replication as they are present in yeast. Rather, "origins" in higher eukaryotes are selected by trans-acting factors as well as by higher order chromatin structure that is likely coordinated with transcription and histone modifications. Unlike the yeasts, defined origins in higher eukaryotes may be the exception and not the rule. Such origin elements

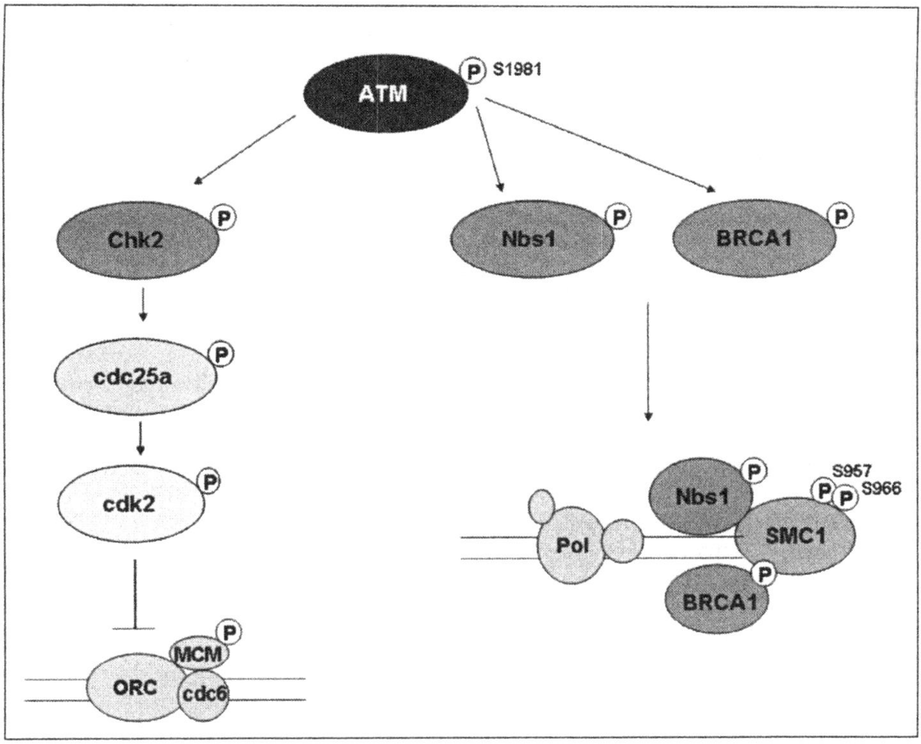

Figure 2. ATM regulates the intra-S-Phase checkpoint by controlling two parallel pathways that suppress DNA replication. In one pathway (left), ATM controls origin activation by inhibiting prereplication complex function through a chk2-cdc25a-cdk2 pathway. The second pathway (right) involves the recruitment of the SMC1 cohesin in a phospho-NBS1 and phospho-BRCA1 dependent manner, blocking the advancement of the replication fork.

are recognized by prereplication (preRC) protein complexes and other downstream factors that recruit and regulate the assembly and/or maintenance of DNA replication forks during S-phase. This includes ORC, Cdc6, the Mcm proteins, CTF4, Cdc45, RPA and the S-Phase kinases Cdk2-cyclin E and Cdc7/Dbf4.[23] During normal DNA replication (in the absence of DNA damage), ATM (and ATR) negatively regulates origin firing through the phosphorylation of Cdk2 and possibly other preRC components such as the MCMs.[22,24] Both ATM and ATR are believed to be activated by feedback from active replicons and this likely occurs through signaling from RPA-ssDNA. In *Xenopus* cell-free extracts, both ATM and ATR promote the Mre11-dependent restart of stalled/collapsed replication forks, an event that prevents the accumulation of DSBs.[25] Thus, the transient generation of DSBs during DNA replication may trigger the local activation of ATM on replicating chromatin. By generating transient DSBs during DNA replication, ATM activation would result in the phosphorylation and inactivation of preRC components. Such a prevention of neighboring origin firing would therefore ensure a regulated spacing of origin firing in space and time and ensure that the chromosome is replicated once per cell cycle.

Artemis and the Role of ATM in the G_2/M Transition

In addition to regulating the G_1/S and intra S-phase checkpoints, ATM plays a role in controlling the G_2/M transition. It is well established that the Cdk1/cyclin B complex is directly regulated by Wee1/Myt1 inhibitory kinases and antagonistic Cdc25 phosphatases. After DNA damage, cells normally recover from G_2 arrest via regulation of Cdk1/cyclin B activation and nuclear import.

Additional factors such as the Plk1 kinase, which targets Wee1 for degradation by the 26S proteosome, contribute to the maintenance of the G_2/M transition by regulating the phosphorylation state of Cdk1. Although the details of how ATM regulates the G_2/M transition have yet to be fully understood, recent evidence has implicated the ATM-dependent phosphorylation of artemis at four sites, S516, S534, S538 and S645, in response to IR as key events in this process.[26,27] Artemis, a NHEJ protein, is a member of the SNM1 family of proteins and appears to play multiple cellular roles in maintaining genomic stability. Artemis-deficient cells are radiosensitive and lead to Severe Combined Immune Deficiency (SCID) syndrome due to a defect in V(D)J recombination (see below). This may be related to a posited role for Artemis in the repair of a subset of DSBs that are recalcitrant to efficient DNA repair.[28] Moreover, recent work from Legerski et al has also implicated Artemis in the G_2/M checkpoint arrest recovery after DNA damage.[26,27,29] Importantly, Artemis mutants deficient in the ATM-dependent phosphorylation at S516 and S645 enhance the accumulation of Cdk1/cyclin B complexes. Because siRNA-mediated knockdown of Artemis shows an accelerated recovery from G_2/M after treatment with IR, the observed phenotypes with artemis-phosphorylation mutants do not appear to be a result of the failure to repair DNA. Thus, ATM controls the G_2/M transition by maintaining the levels of Cdk1/cyclin complexes through the regulation of Artemis, a negative regulator of G_2/M checkpoint recovery. There are, however, conflicting reports on the role that Artemis may play in enforcing the G_2/M checkpoint. Lobrich and Jeggo have reported that Artemis deficient cells display a proper G_2/M cell-cycle checkpoint, with prolonged arrest in G2.[29] One reason for the discrepancies may be the choice of model systems used in each report. In the experiments performed by Jeggo and Lobrich primary Artemis-deficient cells were used, while in the experiments performed by Legerski and colleagues, well established transformed cell lines were transfected with small-interfering RNAs directed towards Artemis creating a "knock-down" effect.

Activation of ATM Kinase: The Role of the MRN Complex and Autophosphorylation

Given the important biological roles that ATM plays in diverse cellular activities, it is not surprising that understanding its mechanism of activation has been the subject of intense investigation. Although the details of ATM activation have yet to be completely worked out, there have been some exciting, yet controversial findings in the field, especially with respect to the role that ATM autophosphorylation plays in this process. Both the status of the MRN complex (consisting of core components Mre11, Rad50 and Nbs1 [Xrs2 in yeast = MRX]) and ATM autophosphorylation have been described as important events in kinase activation.[30-36] Because agents that alter chromatin structure (chloroquine, NaCl and histone deacetylase inhibitors), without apparently introducing DNA breaks, activate ATM, Bakkenist and Kastan initially suggested that ATM activation occurs as a function of changes in chromatin structure and not necessarily in response to DSBs per se.[37] Importantly, ATM activation results from its conversion from an inactive dimer (or oligomer) to active monomers and this appears to be accompanied by autophosphorylation at S1981.[37] Introduction of an S1981A mutant of ATM into HeLa cells inhibited the IR-inducible intra S-phase and G_2/M checkpoints, but failed to complement AT cells. Bakkenist and Kastan proposed that in the absence of genotoxic stress, the cellular activity of one ATM kinase molecule is hindered through its intermolecular association with a second ATM molecule (Fig. 1), a result that can explain the previous and often observed dominant negative behavior of ATM-kinase deficient protein. However, upon DNA damage or perturbations in chromatin structure, an intermolecular, trans-autophosphorylation event at S1981 was posited to generate active ATM monomers. Although no reports have contradicted the importance for ATM monomerization in its activation, additional reports subsequent to the Bakkenist and Kastan paper have challenged the notion that autophosphorylation at S1981 is causal for its activation.[30-33,38] This is discussed below.

Activation of ATM, at least towards some of its substrates, appears to be linked to the functional status of the MRN complex, indicating that MRN plays a major role in controlling ATM function.[34-36,39-41] Additionally, in the absence of functional MRN, one report has suggested that the p53-binding protein 1 (53BP1) contributes to ATM activation as measured by autophosphorylation at S1981 in *Nbs1*-deficient cells treated with siRNAs directed against 53BP1.[42] Numerous studies in various organisms have shown that the MRN complex participates in several aspects of DSB repair including NHEJ and the facilitation of recombination between sister chromatids.[43] Structural studies suggest that MRN bridges DNA structures via a conserved "hook" motif within the Rad50 coiled-coil domain.[44] Thus, MRN is well suited to be a DNA damage "sensor". However, as Mre11 also exhibits nuclease activity and has the additional capacity to evict nucleosomes at the sites of the breaks, MRN has additional roles in DNA repair besides regulating ATM function.[39,45,46] Although ATM has the intrinsic capacity to bind DNA, MRN, as well as Brca1, also plays a role in recruiting the kinase to free DNA ends.[40] This appears to be mediated by a C-terminal, region of Nbs1.[47] Additionally, using an I-PpoI-induced DSB system, Kastan and colleagues have recently used chromatin immunoprecipitation (ChIP) to confirm that Nbs1 recruits ATM to DSBs.[39]

Lee and Paull initially reported in a seminal paper that, in vitro, MRN apparently stimulates ATM kinase activity towards its substrates Chk2, p53 and the histone variant H2AX.[32,33] Here, protein-protein interactions between ATM and MRN appear to stabilize the association between ATM and its substrates.[33] Additionally, this report showed that kinase-defective ATM inhibited wild type ATM phosphorylation of Chk2, a result that is consistent with the dominant-negative effect that has also been observed with ATM in vivo. These initial experiments were performed in the absence of DNA, leading the authors to suggest that DNA and therefore chromatin structure by implication is not absolutely required for the association between ATM, MRN and substrates such as Chk2 and p53.[33] However, this report did not specifically address the multimeric status of ATM;[33] rather, a subsequent report from the same authors showed that in vitro ATM dimers were activated towards Chk2 and p53 in the presence of MRN.[32] Surprisingly, ATM autophosphorylation at S1981 was not required for its dissociation into monomers, a result that apparently contradicts with the original model from Bakkenist and Kastan.[37] Thus, purified MRN complex and DNA also activate ATM towards p53 and this occurred independently of the status of S1981 phosphorylation. Here, the authors proposed that the MRN complex senses DNA breaks, binds and unwinds DNA ends and recruits and dissociates ATM dimers in a manner independent of ATM autophosphorylation at S1981.[33] Thus, in their first report, which failed to examine the relationship between ATM oligomerization and kinase activation, Lee and Paull concluded that DNA was not required for stimulation of ATM activity.[33] However, in a second report,[32] Lee and Paull examined the oligomeric nature of ATM and found that, as reported previously by Bakkenist and Kastan,[37] a dimer to monomer transition was important for ATM activation. Lee and Paull also concluded that free DNA ends, as minimally recognized by the MR subcomplex, were required for the stimulation of ATM kinase activity. The discrepancy in the two reports from the Paull laboratory regarding the role that ATM dimerization plays in ATM activation might be explained (as discussed before[48]) by the fact that the original experiments selected for ATM monomers that were "pre-activated", perhaps as a consequence of the purification itself. Nevertheless, the second Lee and Paul paper,[32] offered biochemical evidence that the MRN heterotrimer both recruits and activates ATM at sites resembling DNA damage. These results conflict with a more recent report that used an in vivo ChIP-based assay to determine that unphosphorylated (S1981) and inactive ATM was not detected near I-PpoI induced breaks.[39] The reason for this discrepancy is not clear, although it is important to mention that the Lee and Paull experiments were not performed in a native setting as they were done in vitro with recombinant proteins and plasmid DNA devoid of chromatin structure, an important determinant in regulating ATM function.

The Role of ATM Autophosphorylation: Discrimination between Kinase Activation and Activity

It is clear that the transition from a dimeric to monomeric form is important for ATM activation, but additional reports have also questioned the role of S1981 phosphorylation as a causative agent in ATM activation.[30,38] For example, using a *Xenopus* cell-free system, Gautier and colleagues observed the stoichiometric conversion of ATM dimers into monomers occurred independently of S1981 autophosphorylation.[30] These authors proposed that ATM activation occurs in two distinct steps: (1) dimeric ATM is recruited to DSBs and dissociates into monomers in a manner facilitated by MRN and (2) Nbs1 activity is required to convert unphosphorylated and inactive ATM monomers to active monomers in a step that does not require DNA. A further examination of the role of ATM autophosphorylation in its role in kinase activation utilized a transgenic mouse model whose sole source of ATM was derived from a mutant allele incapable of phosphorylation at S1987, the apparent murine equivalent of human S1981.[38] In this mouse model, ATM-dependent functions were found to be normal at the cellular and organismal level, suggesting that autophosphorylation at S1987 is a consequence, rather than a cause of ATM activation.[38] However, S367 and S1983 are also autophosphorylated and mutations at these sites render cells defective in ATM activation and fail to correct the radiosensitivity and the G_2/M checkpoint in AT cells.[49] Whether autophosphorylation at other residues such as S1983 is compensatory for a failure to autophosphorylate at S1981 has not been determined. Regardless of the mechanism, it does appear that S-1981P is a good index of ATM activation and with all of the attention given on the role of ATM autophosphorylation in its activation as a kinase, it is also important to note that acetylation of ATM by Tip60 and MOF as well as PP2A and PP5 phosphatases have been also implicated in ATM activation.[50-53] Thus, additional signals and events outside of phosphorylation likely participates in regulating ATM function.

Although recruitment of ATM to DNA is important for its activation, it is not however sufficient for this process. Indeed, the C-terminal motif of Nbs1 has been shown to recruit ATM to sites of DNA damage,[47] nevertheless subsequent report that mice carrying a truncated, hypomorphic allele in which the C-terminal 24 residues have been truncated ($Nbs1^{\Delta C/\Delta C}$ and representing those residues that recruit ATM to sites of DNA damage), show some normal features of ATM activation.[41] Additionally, $Nbs1^{\Delta C/\Delta C}$ embryonic fibroblasts display normal G_1/S checkpoints, but they show a reduced IR-dependent phosphorylation of the cohesion Smc1 and the pro-apoptotic protein BID. These observations led the authors to conclude that the C-terminus of Nbs1 governs the access of activated ATM to its substrate Smc1 and is consistent with a previous report that described the ATM-Nbs1-Brca1-dependent phosphorylation of Smc1.[40] In contrast to the mild checkpoint defects observed, $Nbs1^{\Delta C/\Delta C}$mice possess significant apoptotic defects. Clearly, the C-terminus of Nbs1 is an important determinant in ATM activity in vivo, but important functions such as IR-dependent phosphorylation of p53 occur in its absence. Additionally, Kastan et al have clearly shown that functional Nbs1 is not absolutely required for ATM activation although it is required for the localization of activated ATM to irradiation induced foci (IRIF).[39] Thus, although the MRN complex appears to occupy an important position in regulating ATM function, its requirement is certainly not absolute for ATM kinase activity. Therefore, as discussed by Kastan et al,[39,40] it is important to distinguish between ATM "activation" and "activity" as they are distinct cellular read outs. Thus, ATM substrate phosphorylation is not necessarily a measurement of kinase activation. This is because activated ATM may have activity towards some, but not all, of its substrates and this depends on the context of the components of the MRN complex. Thus, the C-terminus of Nbs1, in conjunction with other components of MRN and additional factors as discussed below, indicate that distinct molecular factors determine ATM signaling specificity in response to various forms of DNA damage.

Modulation of ATM Function in Response to Various Types of Genotoxic Stress

In addition to MRN, ATM binds other cellular factors, some of which modulate ATM activity as a function of the type of DNA damage (i.e., irradiation induction vs. replication stress). As revealed in *Xenopus* cell-free extracts,[5] ATM is necessary for the ATR-Chk1 pathway in response to DSBs, but apparently not in response to the replication inhibitor, hydroxyurea and this appears to be mediated by ATM-dependent phosphorylation of TopBP1, an ATM activating protein.[5] Recently, Kanu and Behrens[31] have identified an ATM interacting protein (ATMIN) as a factor that, like Nbs1, binds to the C-terminus of ATM. ATMIN appears to control the activity of ATM in the absence of

DSBs as it is required for ATM activity in response to hypotonic stress and HU. In contrast, Nbs1 is required for ATM/ATMIN dissociation after IR. Thus, competition for binding to ATM between Nbs1 and ATMIN represents the existence of a class of cofactors that regulate ATM function in a context-dependent manner. Consistently, ATM activation in response to hypotonic stress or inhibitors of DNA replication is normal in Nbs1-deficient mice.[54] ATMIN–/– cells do not display radiosensitivity and have a normal G_2/M checkpoint, but are sensitive to chloroquine, a drug that alters chromatin structure without the introduction of DSBs.[31] Intriguingly, loss of ATMIN function reduced ATM autophosphorylation at S1981/1987 in response to various stimuli. However, phosphorylation of the ATM substrate Smc1 occurred essentially normally in ATMIN–/– cells, further demonstrating that autophosphorylation at S1981/1987 is not an absolute requirement for at least a subset of ATM substrate phosphorylation events. Thus, differential binding to ATM during various types of DNA damage responses may control its function. This phenomenon may be further exemplified through studies between ATM and TRF2 in the DDR elicited in response to dysfunctional telomeres (Fig. 3). When telomere function is perturbed through replicative attrition in telomerase-deficient cells or through inhibition of the end-protecting shelterin complex (Fig. 3), cells exhibit many hallmark features of ATM signaling, including autophosphorylation and activation of the two pathways. In contrast, perturbation of shelterin through depletion of either Pot1a or Pot1b, a ssDNA-binding protein, primarily activates the ATR signaling pathway in response to telomere dysfunction.[6,7] Importantly, compromising the function of the shelterin component TRF2 through a dominant negative strategy elicits a robust DNA damage signal that can be cytologically detected as "telomere-dysfunction induced foci" (TIFs). Here, ATM, but

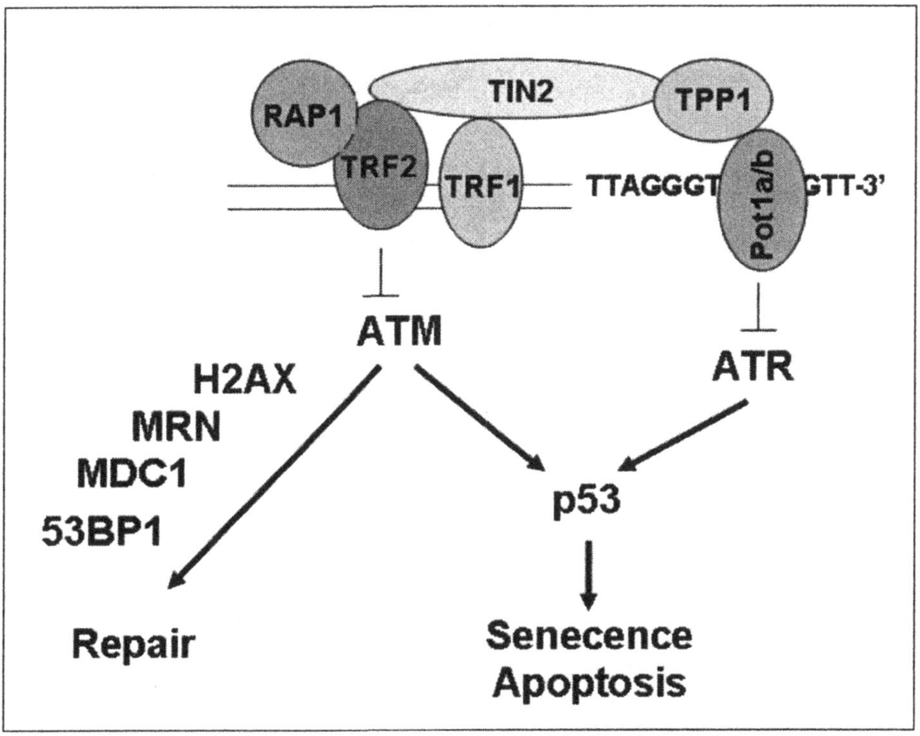

Figure 3. The shelterin complex represses the DSB response at telomeres through the inhibition of both ATM and ATR. Pot1a/1b represses ATR which prevents the activation of Chk1 and p53. Trf2 represses ATM which is needed to prevent the accumulation of chromosomal breaks repaired by NHEJ.

apparently not ATR, is activated, autophosphorylated at S1981 and this leads to cell cycle arrest. Thus, TRF2 specifically represses ATM function at telomere ends and because TRF2 possesses no known function outside of telomeres, this component of shelterin may represent a unique mode of regulating ATM function.[6,7,55] Although TRF2 binds ATM and negatively regulates its function,[55] how this shelterin component prevents ATM activation is unclear. Thus, it will be interesting to see if this depends on the status of the MRN complex.

ATM: Coordinating DSB Repair and Chromatin Structure

Various lines of evidence suggest that in response to DBSs, chromatin adopts a relaxed structure.[56] How this is accomplished and how this chromatin compares in structure and composition to other forms of "relaxed" chromatin (i.e., at active euchromatin) is unknown. Clearly, upon DNA damage the local chromatin structure at and near the physical site of the break is altered (Fig. 4). In

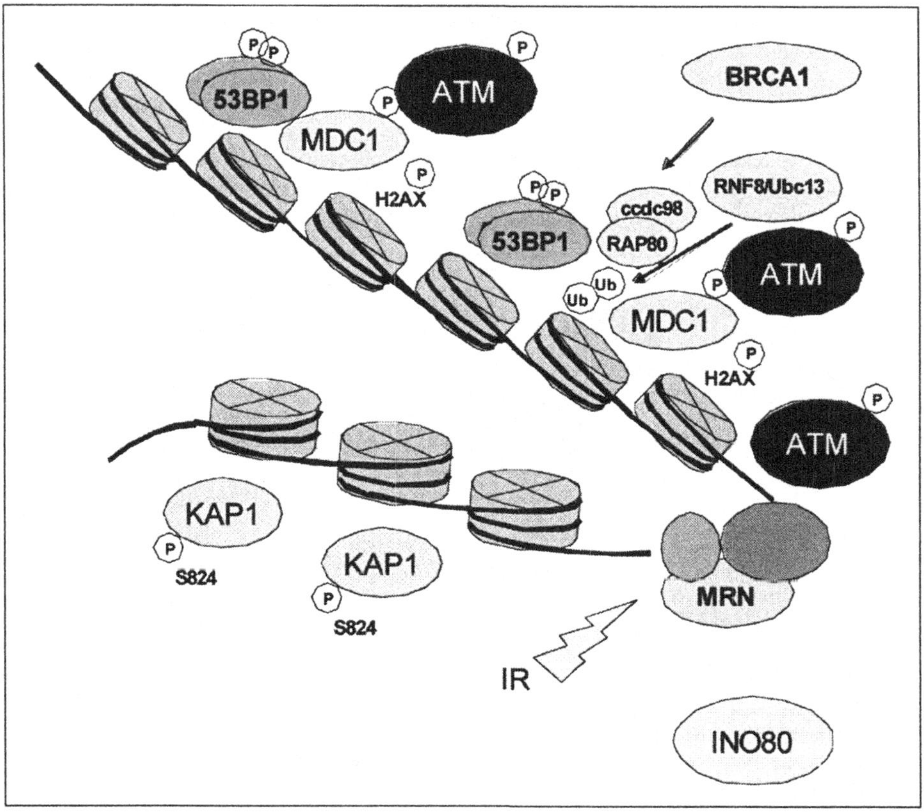

Figure 4. In response to DSBs, ATM accumulates at DNA repair "foci". ATM phosphorylates the C-terminal tail of the histone variant H2AX generating γ-H2AX which leads to the direct binding of Mdc1 and subsequent recruitment of DNA repair factors such as 53BP1 and Brca1. ATM-dependent phosphorylation of Mdc1 recruits the RNF8/Ubc13 ubiquiitylase near DSBs. Here, ubiquitylated histones H2A and H2AX recruit both 53BP1 and BRCA1 by two independent means. BRCA1 is recruited by ccdc98 (abraxas)/RAP80 and 53BP1 is recruited by various histone modifications including dimethylated histone H4 at lysine 20. Importantly, γ-H2AX accumulates and spreads along the chromatin fiber for up to megabase distances distal to the physical site of DNA damage. Upon DNA damage, the local chromatin structure at and near the physical site of the break is altered by remodeling complexes such as INO80 and ATM-dependent phosphorylation KAP-1.

both yeast and mammals, histones are evicted from the site of the break in a manner that depends upon MRN components as well as other chromatin remodeling factors such as Ino80.[39,45,46,57,58] Moreover, a recent report from Shiloh et al has identified the ATM-dependent phosphorylation of KAP-1 (KRAB-associated protein), a factor implicated in transcriptional corepression, as an important determinant in DSB-induced relaxation of chromatin.[59] Importantly, ATM phosphorylates KAP-1 at S824, leading to the spreading, but not focus formation, of phosphorylated KAP-1 throughout the nucleus. Prevention of this event leads to an impairment in DSB-induced chromatin decondensation and renders cells sensitive to DSB-inducing agents. Interestingly, the generation of DSBs appears to induce a wave of chromatin relaxation that spreads throughout the entire genome and this is mediated by ATM and KAP-1.[59] The relationship between ATM and KAP-1 clearly underscores the intimacy between the regulation of DSB repair and chromatin structure. The mechanisms and targets of Kap-1 are currently unknown, but it will be interesting to learn if and how it is related to posttranslational modifications found on histone tails.

ATM and Its Accumulation at DNA Repair "Foci"

In addition to targeting chromatin regulators like KAP-1, ATM also phosphorylates the C-terminal tail of the histone variant H2AX in response to DSBs, generating γ-H2AX.[60-62] Thus, ATM directly targets the nucleosome itself and therefore participates in a "histone code" (Fig. 4).[63] Phosphorylation of H2AX represents the founding member of a growing list of histone modifications implicated in the coordination of DSB repair and chromatin function.[45,46,64-67] H2AX is one of several histone variants and comprises up to 30% of nucleosomes in some cell types.[68-70] Importantly, seminal work from Bonner et al has shown that in response to DSBs, γ-H2AX accumulates and spreads along the chromatin fiber for up to megabase distances distal to the physical site of DNA damage.[61,62] Such a spreading phenomenon can be readily detected with antibodies as irradiation-induced foci (IRIF). γ-H2AX directly binds to the BRCT protein Mdc1 which, in turn, recruits the DNA repair factors such as 53BP1 and Brca1 to IRIF (Fig. 4).[71-75] As the forkhead (FHA) domain of Mdc1 has the capacity to bind additional ATM molecules,[73] ATM kinase is recruited to loci outside of the immediate vicinity of the DSB where it phosphorylates additional H2AX molecules that recruit additional factors. This can be detected as IRIF. Thus, Mdc1 forms a molecular bridge between H2AX and ATM and this contributes to the propagation of a DNA damage signal and the accumulation and phosphorylation of a host of nearby DSB repair factors.[73] Such a positive feedback loop for ATM function represents a diffusible mechanism for ATM activation.[73] Therefore, both chromatin and soluble-based mechanisms serve to activate and propagate ATM signaling events. Recent evidence has also shown that the RNF8/Ubc13 ubiquitin ligase binds to Mdc1 in a manner dependent upon the ATM phosphorylation of Mdc1 within a TQXF cluster.[76,77] Once bound to Mdc1, RNF8/Ubc13 ubiquitnates H2A and H2AX. These ubiquitynated histones bind ccdc98 (abraxas)/RAP80 which are capable of recruiting BRCA1 to IRIF.[78,79] Interestingly, RNF8 is required for 53BP1 IRIF formation, but RAP80 is not. Therefore, 53BP1 and BRCA1 are each recruited to IRIF through overlapping (γ-H2AX-Mdc1-RNF8) but distinct mechanisms. Unlike BRCA1, acetylation of histone H4 at K16 and methylation of H4-K20 are also required for the localization of 53BP1, as well as its yeast orthologue Crb2, to sites of DNA damage.[80,81] Importantly, the tandem Tudor domain of 53BP1 interacts with H4-K20, preferentially in its dimethylated form.[82] Thus, four separate histone modifications in addition to DSB repair proteins such as Mdc1 are required for 53BP1 IRIF formation. This indicates that 53BP1 forms an intriguing link between DSB repair and chromatin structure/function and this appears to be ATM dependent. A kinetic analysis of DSB repair showed that ATM and other factors that form IRIF, such as H2AX and 53BP1, are required to repair a small subset of DNA breaks that possess structures difficult to repair.[28] Moreover, genetic analysis with murine knock-outs have indicated that animals defective in ATM, H2AX, Mdc1 and 53BP1 all possess overlapping traits including growth retardation, sensitivity to ionizing radiation and immune deficiencies (Table 1). Specifically, these factors have been shown to function in programmed DSB repair events in T- and B-cells of lymphoid origin that utilize "unusual" methods for generating and resolving DSBs.

Table 1. Overlapping traits in animals defective in various genes associated with AT

Gene	Phenotype	IR Sensitive	Tumor Prone	CSR Defect	P53 Cross	References
ATM	T-cell lymphomagenesis within 8 months; TCR locus translocations;	Yes	Yes	Moderately defective	NR	15
p53	Die of T-cell lymphomas with mean age of 5 months; aneuploid; centrosome amplification	Yes	Yes	None	–	13,14
H2AX	Small, growth retarded, sterile males	Yes	Low	Moderately defective	T/B-cell lymphoma	68-70
53BP1	Small, growth retarded, fertile	Yes	Low	Severely defective	T/B-cell lymphoma	91, 92
NHEJ	Impaired lymphocyte development, small, growth retarded, DNA ligase IV is lethal	Yes	Low	ND*	B-cell lymphoma with t(12; 15) translocation	2-4
MDC1	Small, growth retarded	Yes	Low	Mildly defective in IgG1	NR	73

*Since these animals do not perform V(D)J recombination, they are unable to undergo CSR.

Murine ATM Models: Revealing the Mechanisms of ATM in DSB Repair in Lymphocytes

Mice homozygous for mutations in ATM (ATM–/–) recapitulate several in vivo aspects of the AT phenotype.[15,83] This includes growth retardation, immune defects, gonadal abnormalities, sensitivity to X-rays and a high incidence of T-cell lymphomagenesis.[15,83] Moreover, murine ATM–/– cells, as per their human counterparts, are defective in cell cycle checkpoints and are hypersensitive to X-rays, phenotypes that can be explained by defects in the cellular response to DSBs. ATM–/– mice fail to display ataxia or telangiectasia and this is probably due to their relatively short life span as nearly 100% of ATM–/– mice die of T-cell lymphomas within 2-8 months of birth (Table 1). A variety of cytological, biochemical and genetic evidence suggests that ATM interacts with H2AX, 53BP1 and Mdc1 in controlling DNA end joining events that occur during the development of T- and B-cells (Figs. 5A,B). This is most pronounced for class switch recombination (CSR) in B-cells and is discussed below in the context of a novel DNA end joining mechanism that coordinates DSB repair and chromatin function.[28,69,84,85] Such a pathway suppresses the development of translocations and lymphomagenesis.

As ATM plays a major role in the development of lymphocytes, major progress has been made in understanding its function in lymphoid T- and B-cells. During development of the immune system, both T- and B-cells use programmed DSB repair. These NHEJ events to assemble the exons encoding immunoglobulins (Igs) and T-cell receptors (TCR) by V(D)J recombination, a process that generates the variable regions of lymphocyte antigen receptors (Fig. 5A).[2-4] This reaction is initiated during G_1 phase by the action of the RAG endonuclease complex. RAG cleaves DNA in synaptic complexes at defined recombination signal sequences and results in the formation of both coding and signal joints (Fig. 5A). DSBs introduced at both TCR and IgH loci are processed through an unusual, closed hairpin structure that is efficiently resolved through ATM function.[2,3] Through poorly understood processes that likely include chromatin contraction,[86] distal DNA sequences are brought together during synapsis prior to resolution of the DSB.

ATM and V(D)J Recombination

Chromosomal translocations from both human and murine ATM–/– tumors involve the T-cell receptor locus (TCR). Moreover, antigen receptor translocations in ATM–/– mice most frequently involve the TCRα/δ locus.[87] This strongly suggests that V(D)J recombination might be disrupted in the absence of ATM function. Indeed, when ATM–/– mice are bred with RAG deficiency, the resulting RAG–/–/ATM–/– double mutants do not undergo lymphomagenesis.[88] Moreover, ATM is recruited to RAG-induced DSBs during G_1 and the phenotype of both ATM–/– humans and mice imply that ATM performs some function during antigen gene processing. Despite this, however, ATM is not absolutely required for V(D)J recombination. Recent elegant work from Sleckman et al[89] have clarified the role of ATM in the V(D)J mechanism. Here, the authors have shown that ATM functions directly in the repair of chromosomal DSBs through the maintenance of free DNA ends generated during T-lymphocyte development.[89] Although signal joint formation proceeds normally in an ATM–/– pre B-cell line, unrepaired coding ends accumulate in these cells as they are not stabilized in post synaptic complexes.[9] Here, approximately 10-20% of loci in ATM–/– thymocytes or ATM–/– pre B-cells develop persistent, unrepaired coding ends. Such "loose" ends are often aberrantly resolved through the generation of translocations. Indeed, "long lived" peripheral ATM–/– lymphocytes contain unresolved RAG-dependent DSBs, a condition that contributes to increases in the levels of chromosome 12 translocations at the immunoglobulin heavy chain locus (IgH).[90] Therefore, during V(D)J recombination, ATM regulates a checkpoint that prevents the transmission of DSBs through proper synapsis and end joining and this appears to operate independently of its control of the Chk2-p53 pathway in G_1. Clearly, developing T-cells require ATM for proper antigen receptor processing. However, in T-cells ATM may have additional roles outside of DSB repair pathways and this could include JNK signaling.

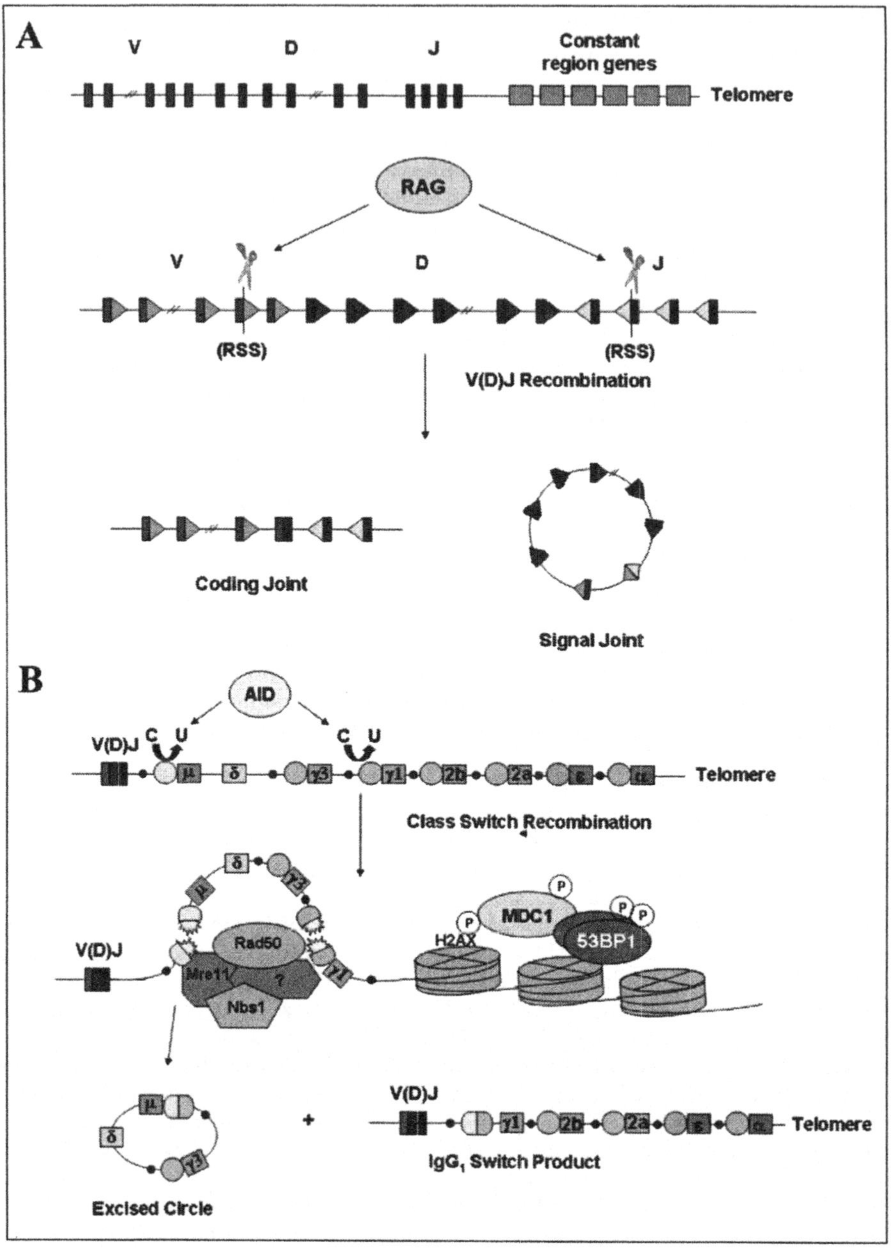

Figure 5. Programmed DSB repair processes in lymphoid cells. A) V(D)J recombination is carried out by the RAG protein complex consisting of RAG-1 and RAG-2. RAG recognizes and induces DNA cleavage at recombination signal sequences (RSS). The broken coding sequence is rejoined via NHEJ and produces a rearranged gene segment. B) Following V(D) J recombination in B-cells, class switch recombination (CSR) produces secondary antibody isotypes. This occurs through AID-dependent introduction of DSBs in specified S-regions. The intervening DNA between the S regions is excised and the free DNA ends are rejoined via NHEJ to link the variable domain with the switched antibody heavy chain.

Class Switch Recombination and ATM: Anchoring DSB Breaks to Suppress Translocations

In response to cytokine stimulation, mature B-cells express different Ig constant regions through the process of class switch recombination (CSR), a DSB-mediated, NHEJ-like process that occurs in the germinal center.[2-4] The end result is the generation of a secondary antibody isotype that occurs through the juxtapositioning of a rearranged V(D)J exon to a "switched" *IgH* constant region (Fig. 5B). By deaminating cytosine to uracil on a template provided through transcription, AID induces DNA repair pathways that result in the formation of DSBs in the switch locus, as the uracil residues are processed into DSB intermediates.[2,4,91] During CSR, multiple AID-derived DSBs occur within S regions, leading to both internal deletions and long-range DNA repair events that join S region DNA fragments. Remarkably, the joining of DNA segments often lie up to 100 kb apart from each other on murine chromosome 12. Thus, V(D)J and CSR represent two distinct DSB repair mechanisms that are each necessary for the generation of antibody diversity (Fig. 5B). CSR requires successful completion of V(D)J recombination, cell proliferation and germline transcription through the *IgH* switch (S) locus on the sub-telomeric portion of murine chromosome 12 (Fig. 5B).[2,4] Failure to properly perform CSR often results in chromosomal translocations, particularly in the context of p53 deficiency.[82,84,92-94] This is important because B-cell lymphomas, which represent the vast majority of lymphomas diagnosed in the Western world, often possess translocations involving IgH fused to oncogenes such as c-myc. Thus, the mechanisms that regulate V(D)J and CSR are highly relevant to understanding lymphomagenesis. CSR is initiated by the activation induced deaminase (AID).[2,4,91]

The signaling events emanating from programmed DSBs at IgH bear many hallmark features of the DDR [2-4]. Indeed, ATM and its associated factors that form IRIF (i.e., H2XA, Mdc1, 53 BP1, MRN and ATM) have each been shown to function in CSR.[4,68,70,73,94-96] What is the role of ATM and these related factors in CSR at IgH? Defective CSR in ATM$^{-/-}$ B-cells is not due to alterations in switch region transcription, accessibility, or recruitment of factors to IRIF.[97] Rather, only long-range switch recombination is defective, a scenario that has also been found in 53 BP1-defective B-cells. Thus, ATM, 53 BP1 and H2XA-defective B-cells cannot efficiently repair AID-induced breaks at IgH and this appears to be due to a defect at the level of DNA recombination.[69,95] However, unlike ATM$^{-/-}$ animals, mice defective in 53 BP1 and H2XA are at best mildly tumor prone (Table 1). This is because, in contrast to ATM, deficiencies in H2XA and 53 BP1-defective mice do not significantly impact p53 function. Therefore, p53 suppresses the outgrowth of cells harboring 53 BP1-driven translocations. Indeed, in the context of p53 deficiency, the double-mutant animals rapidly develop both T- and B-cell lymphomas, in a manner distinct from p53-mediated lymphomagenesis in both timing and tumor spectrum.[84,94,98] Notably for B-cell lymphomas, these tumors possess clonal translocations that involve AID-dependent, IgH/c-myc fusions.[70,92,94,98] As deregulation of c-myc creates "oncogenic" stress, this may also activate the p53 pathway through the ARF tumor suppressor.[92] Thus, AID-induced DSBs at IgH activate an ATM-dependent pathway that requires the cooperation of H2XA, Mdc1 and 53 BP1 in suppressing translocations through the promotion of efficient end joining. An analogous pathway may operate at dysfunctional telomeres (Fig. 3). Because both the AID-dependent initiation and resolution of DSBs (in those cases where the breaks are resolved), occurs normally in ATM$^{-/-}$, 53 BP1$^{-/-}$ and H2XA$^{-/-}$ B-cells, ATM regulates synapsis over fairly large distances at IgH. Other kinases such as ATR and DNA-PK may also participate in this process.[95] The mechanisms regarding this are unknown, but through the coordinated processes of DSB repair and maintenance of chromatin structure and function, ATM appears to "anchor" DSBs,[84] a process that suppresses chromosomal translocations at IgH.

Acknowledgements

We would like to thank Julio Morales for comments on the manuscript. Research in the Carpenter laboratory was supported by the NIH and Welch (AU-1569).

References

1. Savitsky K, Bar-Shira A, Gilad S et al. A single ataxia telangiectasia gene with a product similar to PI-3 kinase. Science 1995; 268(5218):1749-1753.
2. Dudley DD, Chaudhuri J, Bassing CH et al. Mechanism and control of V(D)J recombination versus class switch recombination: similarities and differences. Adv Immunol 2005; 86:43-112.
3. Jones JM, Gellert M. The taming of a transposon: V(D)J recombination and the immune system. Immunol Rev 2004; 200:233-248.
4. Kenter AL. Class switch recombination: an emerging mechanism. Curr Top Microbiol Immunol 2005; 290:171-199.
5. Yoo HY, Kumagai A, Shevchenko A et al. Ataxia telangiectasia mutated (ATM)-dependent activation of ATR occurs through phosphorylation of TopBP1 by ATM. J Biol Chem 2007; 282(24):17501-17506.
6. Denchi EL, de Lange T. Protection of telomeres through independent control of ATM and ATR by TRF2 and POT1. Nature 2007; 448(7157):1068-1071.
7. Guo X, Deng Y, Lin Y et al. Dysfunctional telomeres activate an ATM-ATR-dependent DNA damage response to suppress tumorigenesis. EMBO J 2007; 26(22):4709-4719.
8. O'Driscoll M, Ruiz-Perez VL, Woods CG et al. A splicing mutation affecting expression of ataxia telangiectasia and Rad3-related protein (ATR) results in Seckel syndrome. Nat Genet 2003; 33(4):497-501.
9. Matsuoka S, Ballif BA, Smogorzewska A et al. ATM and ATR substrate analysis reveals extensive protein networks responsive to DNA damage. Science 2007; 316(5828):1160-1166.
10. Kastan MB, Zhan Q, el-Deiry WS et al. A mammalian cell cycle checkpoint pathway utilizing p53 and GADD45 is defective in ataxia telangiectasia. Cell 1992; 71(4):587-597.
11. Chehab NH, Malikzay A, Appel M et al. Chk2/hCds1 functions as a DNA damage checkpoint in G(1) by stabilizing p53. Genes Dev 2000; 14(3):278-288.
12. Hirao A, Kong YY, Matsuoka S et al. DNA damage-induced activation of p53 by the checkpoint kinase Chk2. Science 2000; 287(5459):1824-1827.
13. Donehower LA, Harvey M, Slagle BL et al. Mice deficient for p53 are developmentally normal but susceptible to spontaneous tumours. Nature 1992; 356(6366):215-221.
14. Jacks T, Remington L, Williams BO et al. Tumor spectrum analysis in p53-mutant mice. Curr Biol 1994; 4(1):1-7.
15. Xu Y, Ashley T, Brainerd EE et al. Targeted disruption of ATM leads to growth retardation, chromosomal fragmentation during meiosis, immune defects and thymic lymphoma. Genes Dev 1996; 10(19):2411-2422.
16. Painter RB, Young BR. Radiosensitivity in ataxia telangiectasia: a new explanation. Proc Natl Acad Sci USA 1980; 77(12):7315-7317.
17. Falck J, Mailand N, Syljuasen RG et al. The ATM-Chk2-Cdc25A checkpoint pathway guards against radioresistant DNA synthesis. Nature 2001; 410(6830):842-847.
18. Falck J, Petrini JH, Williams BR et al. The DNA damage-dependent intra-S-Phase checkpoint is regulated by parallel pathways. Nat Genet 2002; 30(3):290-294.
19. Lim DS, Kim ST, Xu B et al. ATM phosphorylates p95/nbs1 in an S-phase checkpoint pathway. Nature 2000; 404(6778):613-617.
20. Kim ST, Xu B, Kastan MB. Involvement of the cohesin protein, Smc1, in Atm-dependent and independent responses to DNA damage. Genes Dev 2002; 16(5):560-570.
21. Yazdi PT, Wang Y, Zhao S et al. SMC1 is a downstream effector in the ATM/NBS1 branch of the human S-phase checkpoint. Genes Dev 2002; 16(5):571-582.
22. Shechter D, Gautier J. ATM and ATR check in on origins: a dynamic model for origin selection and activation. Cell Cycle 2005; 4(2):235-238.
23. Zhu W, Ukomadu C, Jha S et al. Mcm10 and And-1/CTF4 recruit DNA polymerase alpha to chromatin for initiation of DNA replication. Genes Dev 2007; 21(18):2288-2299.
24. Cortez D, Glick G, Elledge SJ. Minichromosome maintenance proteins are direct targets of the ATM and ATR checkpoint kinases. Proc Natl Acad Sci USA 2004; 101(27):10078-10083.
25. Trenz K, Smith E, Smith S et al. ATM and ATR promote Mre11 dependent restart of collapsed replication forks and prevent accumulation of DNA breaks. EMBO J 2006; 25(8):1764-1774.
26. Geng L, Zhang X, Zheng S et al. Artemis links ATM to G2/M checkpoint recovery via regulation of Cdk1-cyclin B. Mol Cell Biol 2007; 27(7):2625-2635.
27. Zhang X, Succi J, Feng Z et al. Artemis is a phosphorylation target of ATM and ATR and is involved in the G2/M DNA damage checkpoint response. Mol Cell Biol 2004; 24(20):9207-9220.
28. Riballo E, Kuhne M, Rief N et al. A pathway of double-strand break rejoining dependent upon ATM, Artemis and proteins locating to gamma-H2AX foci. Mol Cell 2004; 16(5):715-724.

29. Jeggo PA, Lobrich M. Artemis links ATM to double strand break rejoining. Cell Cycle 2005; 4(3):359-362.
30. Dupre A, Boyer-Chatenet L, Gautier J. Two-step activation of ATM by DNA and the Mre11-Rad50-Nbs1 complex. Nat Struct Mol Biol 2006; 13(5):451-457.
31. Kanu N, Behrens A. ATMIN defines an NBS1-independent pathway of ATM signalling. EMBO J 2007; 26(12):2933-2941.
32. Lee JH, Paull TT. ATM activation by DNA double-strand breaks through the Mre11-Rad50-Nbs1 complex. Science 2005; 308(5721):551-554.
33. Lee JH, Paull TT. Direct activation of the ATM protein kinase by the Mre11/Rad50/Nbs1 complex. Science 2004; 304(5667):93-96.
34. Morales M, Theunissen JW, Kim CF et al. The Rad50S allele promotes ATM-dependent DNA damage responses and suppresses ATM deficiency: implications for the Mre11 complex as a DNA damage sensor. Genes Dev 2005; 19(24):3043-3054.
35. Uziel T, Lerenthal Y, Moyal L et al. Requirement of the MRN complex for ATM activation by DNA damage. EMBO J 2003; 22(20):5612-5621.
36. You Z, Chahwan C, Bailis J et al. ATM activation and its recruitment to damaged DNA require binding to the C terminus of Nbs1. Mol Cell Biol 2005; 25(13):5363-5379.
37. Bakkenist CJ, Kastan MB. DNA damage activates ATM through intermolecular autophosphorylation and dimer dissociation. Nature 2003; 421(6922):499-506.
38. Pellegrini M, Celeste A, Difilippantonio S et al. Autophosphorylation at serine 1987 is dispensable for murine Atm activation in vivo. Nature 2006; 443(7108):222-225.
39. Berkovich E, Monnat RJ Jr, Kastan MB. Roles of ATM and NBS1 in chromatin structure modulation and DNA double-strand break repair. Nat Cell Biol 2007; 9(6):683-690.
40. Kitagawa R, Bakkenist CJ, McKinnon PJ et al. Phosphorylation of SMC1 is a critical downstream event in the ATM-NBS1-BRCA1 pathway. Genes Dev 2004; 18(12):1423-1438.
41. Stracker TH, Morales M, Couto SS et al. The carboxy terminus of NBS1 is required for induction of apoptosis by the MRE11 complex. Nature 2007; 447(7141):218-221.
42. Mochan TA, Venere M, DiTullio RA Jr et al. 53BP1 and NFBD1/MDC1-Nbs1 function in parallel interacting pathways activating ataxia telangiectasia mutated (ATM) in response to DNA damage. Cancer Res 2003; 63(24):8586-8591.
43. Cherry SM, Adelman CA, Theunissen JW et al. The Mre11 complex influences DNA repair, synapsis and crossing over in murine meiosis. Curr Biol 2007; 17(4):373-378.
44. Wiltzius JJ, Hohl M, Fleming JC et al. The Rad50 hook domain is a critical determinant of Mre11 complex functions. Nat Struct Mol Biol 2005; 12(5):403-407.
45. Morrison AJ, Highland J, Krogan NJ et al. INO80 and gamma-H2AX interaction links ATP-dependent chromatin remodeling to DNA damage repair. Cell 2004; 119(6):767-775.
46. Tsukuda T, Fleming AB, Nickoloff JA et al. Chromatin remodelling at a DNA double-strand break site in Saccharomyces cerevisiae. Nature 2005; 438(7066):379-383.
47. Falck J, Coates J, Jackson SP. Conserved modes of recruitment of ATM, ATR and DNA-PKcs to sites of DNA damage. Nature 2005; 434(7033):605-611.
48. Abraham, RT, Tibbetts, RS. Cell biology. Guiding ATM to broken DNA. Science 2005; 308(5721):510-511.
49. Kozlov SV, Graham ME, Peng, C et al. Involvement of novel autophosphorylation sites in ATM activation. EMBO J 2006; 25(15):3504-3514.
50. Ali A, Zhang J, Bao S et al. Requirement of protein phosphatase 5 in DNA-damage-induced ATM activation. Genes Dev 2004; 18(3):249-254.
51. Goodarzi AA, Jonnalagadda JC, Douglas P et al. Autophosphorylation of ataxia telangiectasia mutated is regulated by protein phosphatase 2A. EMBO J 2004; 23(22):4451-4461.
52. Gupta A, Sharma GG, Young CS et al. Involvement of human MOF in ATM function. Mol Cell Biol 2005; 25(12):5292-5305.
53. Sun Y, Jiang X, Chen S et al. A role for the Tip60 histone acetyltransferase in the acetylation and activation of ATM. Proc Natl Acad Sci USA 2005; 102(37):13182-13187.
54. Difilippantonio S, Celeste A, Fernandez-Capetillo O et al. Role of Nbs1 in the activation of the Atm kinase revealed in humanized mouse models. Nat Cell Biol 2005; 7(7):675-685.
55. Karlseder J, Hoke K, Mirzoeva OK et al. The telomeric protein TRF2 binds the ATM kinase and can inhibit the ATM-dependent DNA damage response. PLoS Biol 2004; 2(8):E240.
56. Kruhlak MJ, Celeste A, Dellaire G et al. Changes in chromatin structure and mobility in living cells at sites of DNA double-strand breaks. J Cell Biol 2006; 172(6):823-834.
57. Morrison AJ, Shen X. DNA repair in the context of chromatin. Cell Cycle 2005; 4(4):568-571.
58. Unal E, Arbel-Eden A, Sattler U et al. DNA damage response pathway uses histone modification to assemble a double-strand break-specific cohesin domain. Mol Cell 2004; 16(6):991-1002.

59. Ziv Y, Bielopolski D, Galanty Y et al. Chromatin relaxation in response to DNA double-strand breaks is modulated by a novel ATM- and KAP-1 dependent pathway. Nat Cell Biol 2006; 8(8):870-876.
60. Burma S, Chen BP, Murphy M et al. ATM phosphorylates histone H2AX in response to DNA double-strand breaks. J Biol Chem 2001; 276(45):42462-42467.
61. Rogakou EP, Boon C, Redon C et al. Megabase chromatin domains involved in DNA double-strand breaks in vivo. J Cell Biol 1999; 146(5):905-916.
62. Rogakou EP, Pilch DR, Orr AH et al. DNA double-stranded breaks induce histone H2AX phosphorylation on serine 139. J Biol Chem 1998; 273(10):5858-5868.
63. Turner BM. Cellular memory and the histone code. Cell 2002; 111(3):285-291.
64. Downs JA, Nussenzweig MC, Nussenzweig A. Chromatin dynamics and the preservation of genetic information. Nature 2007; 447(7147):951-958.
65. Loizou JI, Murr R, Finkbeiner MG et al. Epigenetic information in chromatin: the code of entry for DNA repair. Cell Cycle 2006; 5(7):696-701.
66. Shroff R, Arbel-Eden A, Pilch D et al. Distribution and dynamics of chromatin modification induced by a defined DNA double-strand break. Curr Biol 2004; 14(19):1703-1711.
67. Wurtele H, Verreault A. Histone posttranslational modifications and the response to DNA double-strand breaks. Curr Opin Cell Biol 2006; 18(2):137-144.
68. Bassing CH, Chua KF, Sekiguchi J et al. Increased ionizing radiation sensitivity and genomic instability in the absence of histone H2AX. Proc Natl Acad Sci USA 2002; 99(12):8173-8178.
69. Bassing CH, Suh H, Ferguson DO et al. Histone H2AX: a dosage-dependent suppressor of oncogenic translocations and tumors. Cell 2003; 114(3):359-370.
70. Celeste A, Petersen S, Romanienko PJ et al. Genomic instability in mice lacking histone H2AX. Science 2002; 296(5569):922-927.
71. Bekker-Jensen S, Lukas C, Melander F et al. Dynamic assembly and sustained retention of 53BP1 at the sites of DNA damage are controlled by Mdc1/NFBD1. J Cell Biol 2005; 170(2):201-211.
72. Lee MS, Edwards RA, Thede GL et al. Structure of the BRCT repeat domain of MDC1 and its specificity for the free COOH-terminal end of the gamma-H2AX histone tail. J Biol Chem 2005; 280(37):32053-32056.
73. Lou Z, Minter-Dykhouse K, Franco S et al. MDC1 maintains genomic stability by participating in the amplification of ATM-dependent DNA damage signals. Mol Cell 2006; 21(2):187-200.
74. Stewart GS, Wang B, Bignell CR et al. MDC1 is a mediator of the mammalian DNA damage checkpoint. Nature 2003; 421(6926):961-966.
75. Stucki M, Clapperton JA, Mohammad D et al. MDC1 directly binds phosphorylated histone H2AX to regulate cellular responses to DNA double-strand breaks. Cell 2005; 123(7):1213-1226.
76. Huen MS, Grant R, Manke I et al. RNF8 Transduces the DNA-Damage Signal via Histone Ubiquitylation and Checkpoint Protein Assembly. Cell 2007; 131(5):901-914.
77. Mailand N, Bekker-Jensen S, Faustrup H et al. RNF8 ubiquitylates histones at DNA double-strand breaks and promotes assembly of repair proteins. Cell 2007; 131(5):887-900.
78. Sobhian B, Shao G, Lilli DR et al. RAP80 targets BRCA1 to specific ubiquitin structures at DNA damage sites. Science 2007; 316(5828):1198-1202.
79. Wang B, Matsuoka S, Ballif BA et al. Abraxas and RAP80 form a BRCA1 protein complex required for the DNA damage response. Science 2007; 316(5828):1194-1198.
80. Murr R, Loizou JI, Yang YG et al. Histone acetylation by Trrap-Tip60 modulates loading of repair proteins and repair of DNA double-strand breaks. Nat Cell Biol 2006; 8(1):91-99.
81. Sanders SL, Portoso M, Mata J et al. Methylation of histone H4 lysine 20 controls recruitment of Crb2 to sites of DNA damage. Cell 2004; 119(5):603-614.
82. Botuyan MV, Lee J, Ward IM et al. Structural basis for the methylation state-specific recognition of histone H4-K20 by 53BP1 and Crb2 in DNA repair. Cell 2006; 127(7):1361-1373.
83. Westphal CH, Rowan S, Schmaltz C et al. ATM and p53 cooperate in apoptosis and suppression of tumorigenesis, but not in resistance to acute radiation toxicity. Nat Genet 1997; 16(4):397-401.
84. Bassing CH, Alt FW. H2AX may function as an anchor to hold broken chromosomal DNA ends in close proximity. Cell Cycle 2004; 3(2):149-153.
85. Posey JE, Brandt VL, Roth DB. Paradigm switching in the germinal center. Nat Immunol 2004; 5(5):476-477.
86. Liu H, Schmidt-Supprian M, Shi Y et al. Yin Yang 1 is a critical regulator of B-cell development. Genes Dev 2007; 21(10):1179-1189.
87. Liyanage M, Weaver Z, Barlow C et al. Abnormal rearrangement within the alpha/delta T-cell receptor locus in lymphomas from Atm-deficient mice. Blood 2000; 96(5):1940-1946.
88. Liao MJ, Van Dyke T. Critical role for Atm in suppressing V(D)J recombination-driven thymic lymphoma. Genes Dev 1999; 13(10):1246-1250.

89. Bredemeyer AL, Sharma GG, Huang CY et al. ATM stabilizes DNA double-strand-break complexes during V(D)J recombination. Nature 2006; 442(7101):466-470.
90. Callen E, Jankovic M, Difilippantonio S et al. ATM prevents the persistence and propagation of chromosome breaks in lymphocytes. Cell 2007; 130(1):63-75.
91. Petersen S, Casellas R, Reina-San-Martin B et al. AID is required to initiate Nbs1/gamma-H2AX focus formation and mutations at sites of class switching. Nature 2001; 414(6864):660-665.
92. Franco S, Gostissa M, Zha S et al. H2AX prevents DNA breaks from progressing to chromosome breaks and translocations. Mol Cell 2006; 21(2):201-214.
93. Morales JC, Franco S, Murphy MM et al. 53BP1 and p53 synergize to suppress genomic instability and lymphomagenesis. Proc Natl Acad Sci USA 2006; 103(9):3310-3315.
94. Ramiro AR, Jankovic M, Callen E et al. Role of genomic instability and p53 in AID-induced c-myc-Igh translocations. Nature 2006; 440(7080):105-109.
95. Manis JP, Morales JC, Xia Z et al. 53BP1 links DNA damage-response pathways to immunoglobulin heavy chain class-switch recombination. Nat Immunol 2004; 5(5):481-487.
96. Ward IM, Reina-San-Martin B, Olaru A et al. 53BP1 is required for class switch recombination. J Cell Biol 2004; 165(4):459-464.
97. Lumsden JM, McCarty T, Petiniot LK et al. Immunoglobulin class switch recombination is impaired in Atm-deficient mice. J Exp Med 2004; 200(9):1111-1121.
98. Ward IM, Difilippantonio S, Minn K et al. 53BP1 cooperates with p53 and functions as a haploinsufficient tumor suppressor in mice. Mol Cell Biol 2005; 25(22):10079-10086.

CHAPTER 5

Protein-Protein Interactions in Ataxia Telangiectasia

Steven M. Shell and Yue Zou*

Introduction

Ataxia telangiectasia (AT) is an early-onset genetic disorder characterized by progressive neurondegeneration, chromosome instability and an extreme sensitivity to DNA damaging agents such as ionizing radiation.[1-3] An autosomal recessive disorder, AT occurs in approximately 1 out of 40,000 births in the United States. However this frequency varies dramatically around the world. AT patients appear normal at birth but by age 2-3 years the disease starts showing the signs and by age 10 complete loss of walking occurs.[1-3] MRI analysis of the cerebellum shows cerebellar atrophy by age 10 and a defect in the differentiation and maturation of Purkinje cells. Additional symptoms appear as the disease progresses including oculocutaneous telangiectasias (spidery veins in the eyes), oculomotor apraxia, dysarthria, immunodeficiency, as well as a predisposition to cancer, particularly lymphoid cancers. At present there is no treatment for AT, though patients do receive therapy for secondary effects of the disease. Death occurs frequently during the teenage years and is generally attributed to early-onset incurable cancers or infection due to immunodeficiency.[1-3]

AT is the result of defects in the gene encoding the protein Ataxia telangiectasia-mutated (ATM), a member of the phosphatidylinositol 3-kinase-related protein kinase (PIKK) family. While a variety of mutations in the *ATM* gene are identified in patients, the overall effect is a loss of ATM protein kinase activity.[1,4] The PIKK kinases belong to the conserved family of serine/threonine protein kinases that contain a domain typical of the lipid kinase phosphatidylinositol 3-kinase (PI3K).[5] The mammalian PIKK family includes five protein kinases: ATM, ATM- and Rad3-related (ATR), hSMG-1, the mammalian target of rapamycin, mTOR (also termed FRAP) and the catalytic subunit of the DNA-dependent protein kinase (DNA-PKcs). A sixth member, transformation/transcription domain-associated protein (TRRAP), is part of the histone acetyltransferase complex and required for Myc-dependent oncogenesis but has no protein kinase activity.[3,5,6]

As part of cellular DNA damage responses, PIKK kinases initiate damage signaling cascades that serve to arrest cell cycle progression and promote DNA damage repair by phosphorylating many downstream substrates such as Chk1, Chk2 and p53. ATM and DNA-PKcs primarily respond to DNA double strand breaks (DSBs) while ATR mainly responds to damages generated by UV exposure and other nonDSB DNA damage. However, both ATR and hSMG-1 have been also found to respond to DSBs, although much slower than ATM and ATR activity has been found to be at least partially dependent on ATM. As a central player in cellular responses to DNA damage, ATM and related kinases function through a vast protein-protein interaction network.[3,5-12] A recent study

*Corresponding Author: Yue Zou—Department of Biochemistry and Molecular Biology, James H. Quillen College of Medicine, East Tennessee State University, Johnson City, Tennessee 37614, USA. Email: zouy@etsu.edu

Molecular Mechanisms of Ataxia Telangiectasia, edited by Shamim I. Ahmad.

has shown approximately 700 proteins are affected by the ATM/ATR damage response pathways in mammalian cells exposed to ionizing radiation (IR).[13] Although not comprehensive, this chapter will focus on the protein-protein interactions that mediate the ATM and related PIKK-dependent DNA damage responses, particularly in regard to DNA double strand breaks.

Sensing DNA Damage

As with all DNA damage responses, recognition of damage sites followed by recruitment of damage-signaling and repair factors are the initial steps. Regarding DSBs, the current model of recognition relies on the MRN complex.[3,8,11,14,15] Human MRN is a heterotrimeric complex composed of Mre11, Rad50 and Nbs1.[16] Initially, the MRN interaction with the damage site is transient and a stable interaction requires the recruitment of ATM.[8,17] In undamaged cells ATM exists as an inactive dimer, but following DNA damage it undergoes autophosphorylation of residues S1981, S367 and S1893 to form active ATM monomers.[18] Studies performed with *Xenopus* egg extracts indicate that ATM is recruited to the damaged site and activated through autophosphorylation in an MRN dependent manner.[19] Although MRN is required for ATM activation, it is unclear whether dimeric ATM binds to the MRN complex prior to autophosphorylation.[14,15,18,20] Also 53BP1 appears to be important to the recruitment of ATM to the damage site. 53BP1 is recruited by interaction with the methylated K79 residue of H3 that becomes exposed at the DNA strand break site.[21] Although the recruitment of 53BP1 to the damage site occurs independent of ATM recruitment, 53BP1 has been shown to interact with ATM and promote ATM activation following IR treatment. However this mechanism is still poorly understood.[17,22]

ATM binds to the MRN complex through interactions of the ATM HEAT repeat element with the C-terminal of Nbs1, followed by phosphorylation of Nbs1 at multiple sites.[14,15] Upon DSB formation, active ATM proceeds to interact with and phosphorylate the C-terminus of histone H2AX generating the phosphorylation form of H2AX (γ-H2AX) and further promoting the recruitment of MDC1.[23-26] MDC1 binds to the damage site through interactions with γ-H2AX (via its BRCT domain) and to ATM through its FHA protein-protein interaction domain (which has also been demonstrated to interact with the MRN complex).[23,27] The ATM-MDC1-γH2AX interaction is believed to induce a cyclic recruitment of activated ATM and MDC1 to the damage site and to expand the H2AX phosphorylation.[27] The assembly of these factors and the subsequent phosphorylation events serve to stabilize the initial sensing of the DNA damage site as well as initiate the downstream ATM responses and DNA repair.[23,27,28]

ATM and Cell Cycle Checkpoints

ATM plays a central role in DNA damage checkpoint activation that is responsible for arresting cell cycle in order to promote repair and genome stability. Activation of cell cycle checkpoints is accomplished via a signaling cascade that includes a series of protein kinases that function to activate a variety of cell cycle inhibitors and transcription factors.[3,4,8,11] The primary target of ATM in the signaling is the checkpoint kinase Chk2.[11,29,30] ATM interacts with and activates Chk2 through phosphorylation of T68 and, unlike the damage recognition factors mentioned previously, it disseminates throughout the nucleus thereby transducing the damage response signal.[17] Immediately following phosphorylation by ATM, Chk2 phosphorylates the Cdc25A phosphatase on multiple serine residues, an event that promotes ubiquitination and proteasome-mediated degradation of Cdc25A.[31,32] Degradation of Cdc25A prevents dephosphorylation of Cdk2, which is required for recruitment of DNA polymerase α. This mechanism provides a rapid cell cycle arrest at the intra-S-Phase and G1/S-Phase interface in response to DNA damage.[33,34]

Both Chk2 and ATM have also been shown to modulate the p53 response pathway via protein phosphorylation events. Each protein is capable of interacting with and phosphorylating p53 at multiple sites in its N-terminal domain.[3,8,11] ATM phosphorylates p53 on serine residues 9, 15, 20 and 46 while Chk2 phosphorylates serine residues 15, 18, 20 and 37.[29,35,36] The cumulative effect of p53 phosphorylation increases stability of p53 and enhances p53-mediated transcription initiation.[37,38] p53 phosphorylation prevents association with Mdm2 ubiquitin ligase and its subsequent ubiquitination. Mdm2 (part of the E3 polyubiquitin ligase complex) serves as a negative regulator of p53 by targeting it for nuclear export and degradation.[22,38,39] ATM also phosphorylates Mdm2 in an event designed to further inhibit its ubiquitin ligase activity. Although still unclear, it is believed that this event also serves to modulate other pathways mediated by the E3 polyubiquitin ligase complex.[40] In addition, ATM is shown to phosphorylate the Mdm2 related protein, MdmX at S367. MdmX also inhibits p53 activity through binding to the p53 N-terminal domain and reducing acetylation of p53.[41,42] Furthermore, phosphorylation of p53 by ATM and Chk2 has been shown to increase p53 acetylation by the histone acetyltransferase CBP/p300.[43] The stabilization and increased acetylation of p53 leads directly to the up-regulation and accumulation of p21, an inhibitor of cyclin dependent kinases for negative regulation of cell cycle progression at G1/S-Phase.[44] The modulation of the p53 pathway is although a slower process than that observed with Cdc25A, leads to a more prolonged and sustainable cell cycle arrest.[45]

ATM and DSB Repair

In addition to its role in DNA damage checkpoints, ATM is also involved in activating DSB repair. DNA double strand breaks in cells are generally repaired by two pathways: (i) Homologous recombination, which occurs primarily during S and G2 phases and (ii) Nonhomologous end joining (NHEJ) which occurs at all stages of cell cycle.[4] NHEJ requires a sequential action of a battery of proteins and is initiated by the DNA-PK complex.[46,47] The DNA-PK complex is a heterotrimer composed of the DNA-PKcs kinase and the dimeric Ku70/Ku80 clamp proteins.[48] The Ku heterodimer has a high affinity for the free DNA ends exposed at a DSB and serves to recruit the PIKK kinase DNA-PKcs to form the DNA-PK complex.[49] This complex formation is a critical step in initiating NHEJ, as DNA-PKcs is not activated until it binds the Ku heterodimer.[46,47] Following IR, DNA-PKcs complex undergoes extensive autophosphorylation primarily at S/TQ PIKK consensus sequences. However, recent reports demonstrate that ATM also phosphorylates DNA-PKcs at the T2609 cluster following IR.[50] Furthermore, phosphorylation of the T2609 cluster was shown to be required for efficient DSB repair as indicated by the persistence of γ-H2AX foci when ATM is absent or inhibited.[50] It is believed that phosphorylation of this site on DNA-PKcs is required for Artemis endonuclease activity, which is required to process certain nonligatable break sites and involved in the G_2/M DNA damage checkpoint response.[51] Interestingly, it has also been reported that Artemis itself is phosphorylated in an ATM dependent manner and directly interacts not only with DNA-PK and ATM, but also with 53BP1, which is recruited early during damage recognition.[52] A proposed model for ATM mediated DSB repair is that, following DSB recognition by MRN, ATM is activated and 53BP1 binds the damage site. DNA-PK is then activated via autophosphorylation and ATM-dependent phosphorylation of the T2609 cluster.[50] In the event of a nonligatable break, Artemis is recruited via interactions with DNA-PK and 53BP1. It is then phosphorylated by ATM to process the DNA ends into substrates for the core NHEJ repair.[52]

ATM and ATR

It is generally accepted that ATR is targeted to sites of stretched single-strand DNA (ssDNA) coated with replication protein A (RPA) through interactions with its regulator, ATR interacting protein(ATRIP).[53] Accumulation of long stretches of ssDNA is believed to be a result of replication stress and thus the ATRIP-RPA interaction may serve to regulate the binding of ATR to the chromatin and acts as a sensor for replication stress.[11,53] In addition, DNA damage and replication stress also stimulate an independent (of ATR) recruitment of the heterotrimeric Rad9-Rad1-Hus1 (9-1-1) complex to damage sites, which appears to be essential for activation of the downstream kinases of ATR such as Chk1.[11,54,55] The 9-1-1 complex forms a clamp structure similar to that of PCNA,[56] and is loaded onto the chromatin by the Rad17/RFC-related clamp loading complex.[57,58] Loading of 9-1-1 complex is facilitated by the presence of RPA at the damage site through direct interaction of Rad9 with both the p70 and p32 subunits.[55] The role of 9-1-1 complex in Chk1 activation appears to stimulate ATR-mediated Chk1 phosphorylation via Rad9 binding to topoisomerase

IIβ binding protein (TopBP1) whose activation domain (AD) binds and activates ATR.[59] Rad9 binds directly to topoisomerase IIβ binding protein (TopBP1) through its C-terminal interactions with the BRCT I and II domains in the N-terminus of TopBP1.[60] TopBP1 binds to the ATR/ATRIP complex via an interaction between ATRIP and a C-terminal domain of TopBP1, located between BRCT domains VI and VII. The association of TopBP1 with ATR/ATRIP appears to be dynamic and transient, but serves to greatly increase the ATR kinase activity.[61]

Following sensing of replication stress ATR phosphorylates the transducer kinase Chk1 in the presence of Claspin mediator protein.[62-64] Claspin is a chromatin binding protein that has been shown to bind Chk1 in *Xenopus* egg extracts and is necessary for efficient ATR-dependent phosphorylation of Chk1 residues S317 and S345.[63] In response to replication stress Claspin, which contains 12 S/TQ motifs, is phosphorylated, though it is unclear whether Claspin is a PIKK target. The Chk1 N-terminal kinase domain has been demonstrated to bind directly to the phosphorylated domains of Claspin.[63] Interestingly, the C-terminus of Chk1 appears to regulate the Claspin-Chk1 interaction. However it remains to be determined how this event occurs.[63] Upon activation, ATR/ATRIP and Chk1 modulate the cell cycle in a mechanism similar to that observed for ATM.[3,6,8,11]

In addition to its roles in damage sensing and cell cycle arrest, ATR has recently been shown to interact directly with the Nucleotide Excision Repair (NER) factor XPA.[65,66] XPA is an indispensable factor for NER and has no other role outside of NER.[67,68] Upon UV-irradiation, ATR phosphorylates XPA on residue S196, which is located in the helix-turn-helix motif of XPA's DNA binding cleft, although efficiency of the phosphorylation is relatively low.[66] ATR also promotes the nuclear import or accumulation of XPA following UV-irradiation.[65] These events provide the mechanisms for a possible direct modulation of NER by ATR. However details of the mechanism(s) remain unclear.[65,66]

Although ATR is thought to respond primarily to UV-induced or nonDSB DNA damage, it has also been found to participate in the DSB response.[11] It is suggested that the ATM pathway may generate regions of RPA-coated ssDNA that can serve as an activation signal for ATR/ATRIP.[69-72] On the other hand, a recent study indicates that phosphorylation of TopBP1 on S1131 in the C-terminal BRCT domain by ATM, though dispensable for UV-induced ATR activation, is critical for proper ATR activation in response to double strand breaks.[73] This observation links DSB sensing and initial ATM activation to the subsequent activation of ATR. In addition, it has recently been found that the Ku heterodimer (part of the DNA-PK complex) may play a role in modulating the ATM-dependent activation of ATR in response to IR. Furthermore it has been demonstrated that, in the absence of Ku, p53 activation (as determined by S18 phosphorylation) persists. Interestingly, this phenomenon was largely due to the ATR-dependent phosphorylation of p53 via Chk1.[74] Although the mechanism of this activation remains unclear, it shows multiple avenues of control and coordination between the ATM and ATR mediated DNA damage responses.

ATR and Chk1 activation serves a similar role to the ATM/Chk2 activation as discussed earlier. ATR targets p53 for phosphorylation at residues S15 and S37 while Chk1 phosphorylates residues S9 and S18.[6] These phosphorylation events overlap with those generated by ATM/Chk2 and serve the same purpose: p53 stabilization and increasing p53-mediate transcription activity.[17] In addition, Chk1 also targets the Cdc25 family of phosphatases much like Chk2, thereby preventing the initiation of DNA synthesis.[3,6,17] It is currently believed that ATR activation, in response to IR, serves to augment the cell cycle arrest generated by the initial ATM response. It is interesting to note in a recent report that ATM functions in the UV-induced DNA damage response as well. In response to UV irradiation ATM is phosphorylated at S1981 in an ATR-dependent manner.[75] Also, this event occurs in the absence of a functional MRN response as demonstrated by use of Nbs1 mutants with C-terminal truncations. ATM appears to be an ATR phosphorylation target in the UV-induced DNA damage response. However it is unclear how this event is governed.[75] Although previously viewed as independent, there is increasing evidence indicating that ATM and ATR function cooperatively to respond to genotoxic insults.

ATM and Cell Survival

The primary function of the ATM pathway in response to DNA damage is to transduce the damage signal and coordinate DNA repair and cell cycle arrest. However, it is becoming apparent that this is not the only role of ATM in promoting cell survival following insult. New studies indicate that ATM can modulate certain pro- and anti-apoptotic factors to determine the fate of a damaged cell.[76] One such factor is NEMO (also known as IKKγ). NEMO is a regulatory subunit of the IkB kinase (IKK) complex that in turn regulates the activity of the NF-κB transcription factor.[77] Typically, NF-κB is sequestered in the cytosol by its negative regulator IκB. Following DNA damage, ATM phosphorylates NEMO inducing its ubiquitination and subsequent nuclear export. Once in the cytosol NEMO activates IKK kinase complex leading to phosphorylation and inactivation of IκB and dissociation from NF-κB. Once released, NF-κB shuttles back to the nucleus and activates genes associated with cell survival.[77] In addition to NF-κB, another transcription factor has also been demonstrated as an ATM target. Ca^{2+}/cAMP response element binding protein (CREB) is a transcription factor linked to various cellular growth pathways. Unlike NF-κB modulation, ATM phosphorylates CREB directly and this event serves to inactivate CREB.[78] ATM-mediated phosphorylation of CREB is also observed in UV-irradiated cells, further indicating the cross talk between the ATM and ATR DNA damage response pathways.[79] A third factor involved in ATM-mediated cell survival is the pro-apoptotic protein BID.[80-82] BID is a member of the "BH3-only" factors in the BCL-2 family and is typically found in the cytosol as a full-length, inactive form. BID is cleaved by caspase-8 during induction of apoptosis and translocates to the mitochondria where it initiates the release of cytochrome c. However, when phosphorylated, BID becomes resistant to caspase-8 cleavage in what is described as a regulatory event.[83] In addition, a small portion of BID protein is located in the nucleus. Nuclear-localized BID has been found to be phosphorylated by ATM in response to DSB-inducing treatment in an event that promotes cell survival.[81] However, Kamer et al, have demonstrated that ATM-mediated phosphorylation of BID which consists of only a small portion of the cellular pool does not affect the role of BID in the apoptotic response though it is required for efficient cell cycle arrest.[80] While the role of BID in DNA damage response is still poorly understood, it does provide yet another layer of complexity to the ATM-mediated DNA damage response.

ATM, ATR and Apoptosis

The primary roles of the ATM and ATR response pathways are to promote cell survival following DNA damage by arresting cell cycle and promoting DNA damage repair. However, when DNA damage accumulates to high levels these response pathways also function to initiate apoptosis to eliminate damaged cells, preventing tumorigenesis and promote tissue homeostasis.[76] The most direct effect of ATM and ATR on the initiation of apoptosis is through their interactions with p53. As described above, p53 is a target of both the ATM and ATR pathways and its initial role is to aid in the arrest of cell cycle. However, when p53 accumulation is increased, it promotes the expression of apoptotic gene targets such as BAX (BCL-2 associated X protein), PUMA (p53-upregulated modulator of apoptosis) and the FAS receptor.[84] Upregulation of these proteins leads to the release of cytochrome c from mitochondria, initiating the apoptotic pathways.[84]

Although p53 activation and accumulation play a major role in genotoxin-induced apoptosis, studies of p53-mutated cell lines and mice have implicated other p53-independent mechanisms for induction of apoptosis as well. Activation of Chk1 and Chk2 has been found to stimulate E2F1 transcription activity, which in turn leads to increased transcription of the p73 gene.[85] Although p73 is a required factor in the p53-mediated apoptosis response pathway it has also been deomonstrated to be pro-apoptotic in the absense of functional p53.[86,87] Much like p53, p73 upregulates transcription of pro-apoptotic genes such as BAX and PUMA as well as the gene encoding NOXA that leads to cytochrome c release.[86] It is interesting to note, however, that p73 is typically overexpressed in tumor cells lacking functional p53.[88]

Although not yet fully understood, studies on caspase-2 indicate another possible p53-independent apoptotic pathway. An interesting aspect of this pathway is that Caspase-2 is the only pro-caspase that is constitutively found in the nucleus.[89,90] Caspase-2 is required for induction of apoptosis in response to DNA damages such as DSBs and UV-induced photolesions.[91] Although its mechanism is still under investigation, Caspase-2 can mediate apoptosis in two ways: (1) via activation of the BCL-2 family of pro-apoptotic proteins which cause release of cytochrome c and (2) via BCL-2 independent stimulation of cytochrome c release. Caspase-2 cleaves the pro-apoptotic factor BID and mediates BAX translocation to induce cytochrome c release.[92-94] BID, as discussed above, is present in the nucleus at low levels and is a target of direct phosphorylation by ATM.[80-82] This interaction may provide a mechanism for DNA damage dependent Caspase-2 mediated apoptosis, although no conclusive link between the two pathways has been elucidated.

ATM and ATMIN, a Novel Regulator Partner

As discussed above, in response to genotoxic insult the inactive ATM homodimers undergo autophosphorylation to release active ATM monomers.[18] Recently, Kanu et al have characterized a novel ATM interaction protein, ATMIN, which appears to regulate the stability and autophosphorylation of ATM protein.[95] ATMIN is an 88kDa protein containing four zinc finger domains, multiple S/TQ PIKK kinase consensus phosphorylation sites and a C-terminal putative PEST ubiquitin-mediated protein degradation domain. ATMIN also contains an ATM interaction domain similar to that found in Nbs1 located near the PEST site in the C-terminal of the protein.[95,96] In the absence of DNA damage, ATMIN interacts with a sub-population of ATM in an event that promotes the mutual stabilization of both proteins (although the stabilization is greater for

Table 1. Protein-protein interactions in ataxia telangiectasia

Protein	Function	References
ATM Interaction Proteins		
ATMIN	ATM binding protein, DNA damage mediator	95,96
Artemis	Nuclease, NHEJ factor	52
BID	Pro-apoptotic factor	80,81
Chk2	DNA damage signal transduction kinase	29,97
Creb	Transcription factor	78,79
DNA-PKcs	PIKK kinase, NHEJ repair factor	50
H2AX	Histone	24-26
Ku	DNA binding clamp complex, NHEJ repair factor	74
MDC1	DNA double strand break sensor	23,26,27
Mdm2	Ubiquitin ligase	40
MdmX	Ubiquitin ligase	41,42
Nbs1	DNA double strand break sensor	14,15,19,20
NEMO	NFκB regulatory protein	77
p53	Transcription factor	29,35,36
53BP1	DNA double strand break sensor	17,21,22
TopBP1	Topoisomerase IIβ binding protein	73
ATR Interaction Partners		
ATRIP	ATR binding protein, damage recognition mediator	53
Chk1	DNA damage signal transduction kinase	62-64
Claspin	Chromatin binding factor	62-64
p53	Transcription factor	3,6,17
RPA	ssDNA binding protein	53
TopBP1	Topoisomerase IIβ binding protein	61
XPA	Nucleotide excision repair factor	65,66

ATMIN most likely due to shielding of its PEST domain by ATM).[95] In addition to enhancing ATM stability, ATMIN plays a stimulatory role in the autophosphorylation of ATM at residues S1981 and S1987 in response to genotoxic stimuli. Following IR treatment, ATMIN promotes ATM autophosphorylation and is rapidly displaced by Nbs1. However, unlike Nbs1, ATMIN is not required for downstream signaling events to occur following IR damage.[95] In contrast, ATMIN is required for ATM signaling in response to DNA damage that elicits Nbs1-independent ATM activation, such as hypotonic shock or UV irradiation.[18,95]

Conclusions

It is overwhelmingly clear that the DNA damage response pathways are indispensable for maintenance of genome integrity as well as for continued homeostasis of organisms. The ability of the damage response kinases ATM and ATR to interact with such a broad range of protein factors (Table 1) is essential for the coordination of an ever-expanding network of pathways responsible for cell cycle control and damage repair as well as determining an individual cell's fate following genotoxic insult. However, a complete understanding of how ATM and ATR mediate the DNA damage response(s) has not been available yet. Continued research into the DNA damage response pathways is required to further refine the details of this complex system which may lead to medical advancements in the detection and treatment of diseases such as ataxia telangiectasia.

Acknowledgement

Part of the studies described in this chapter was supported by National Institutes of Health grant CA86927 to Y.Z.

References

1. Chun HH, Gatti RA. Ataxia telangiectasia, an evolving phenotype. DNA Repair 2004; 3:1187-1196.
2. Crawford TO, Mandir AS, Lefton-Greif MA et al. Quantitative neurologic assessment of ataxia telangiectasia. Neurology 2000; 54:1505-1509.
3. Shiloh Y. The ATM-mediated DNA-damage response: taking shape. Trends Biochem Sci 2006; 31:402-410.
4. Jeggo PA, Lobrich M. Contribution of DNA repair and cell cycle checkpoint arrest to the maintenance of genomic stability. DNA Repair 2006; 5:1192-1198.
5. Abraham RT. PI 3-kinase related kinases: 'big' players in stress-induced signaling pathways. DNA Repair 2004; 3:883-887.
6. Shiloh Y. ATM and related protein kinases: safeguarding genome integrity. Nat Rev 2003; 3:155-168.
7. Abraham RT. The ATM-related kinase, hSMG-1, bridges genome and RNA surveillance pathways. DNA Repair 2004; 3:919-925.
8. Bartek J, Lukas J. DNA damage checkpoints: from initiation to recovery or adaptation. Curr Opin Cell Biol 2007; 19:238-245.
9. Burma S, Chen DJ. Role of DNA-PK in the cellular response to DNA double-strand breaks. DNA Repair 2004; 3:909-918.
10. Helt CE, Cliby WA, Keng PC et al. Ataxia telangiectasia mutated (ATM) and ATM and Rad3-related protein exhibit selective target specificities in response to different forms of DNA damage. J Biol Chem 2005; 280:1186-1192.
11. Sancar A, Lindsey-Boltz LA, Unsal-Kacmaz K et al. Molecular mechanisms of mammalian DNA repair and the DNA damage checkpoints. Annu Rev Biochem 2004; 73:39-85.
12. Shechter D, Costanzo V, Gautier J. Regulation of DNA replication by ATR: signaling in response to DNA intermediates. DNA Repair 2004; 3:901-908.
13. Matsuoka S, Ballif BA, Smogorzewska A et al. ATM and ATR substrate analysis reveals extensive protein networks responsive to DNA damage. Science 2007; 316:1160-1166.
14. Falck J, Coates J, Jackson SP. Conserved modes of recruitment of ATM, ATR and DNA-PKcs to sites of DNA damage. Nature 2005; 434:605-611.
15. You Z, Chahwan C, Bailis J et al. ATM activation and its recruitment to damaged DNA require binding to the C terminus of Nbs1. Mol Cell Biol 2005; 25:5363-5379.
16. Stracker TH, Theunissen JW, Morales M et al. The Mre11 complex and the metabolism of chromosome breaks: the importance of communicating and holding things together. DNA Repair 2004; 3:845-854.
17. Lukas C, Falck J, Bartkova J et al. Distinct spatiotemporal dynamics of mammalian checkpoint regulators induced by DNA damage. Nat Cell Biol 2003; 5:255-260.

18. Bakkenist CJ, Kastan MB. DNA damage activates ATM through intermolecular autophosphorylation and dimer dissociation. Nature 2003; 421:499-506.
19. Dupre A, Boyer-Chatenet L, Gautier J. Two-step activation of ATM by DNA and the Mre11-Rad50-Nbs1 complex. Nat Struct Mol Biol 2006; 13:451-457.
20. Cerosaletti K, Wright J, Concannon P. Active role for nibrin in the kinetics of atm activation. Mol Cell Biol 2006; 26:1691-1699.
21. Huyen Y, Zgheib O, Ditullio RA Jr et al. Methylated lysine 79 of histone H3 targets 53BP1 to DNA double-strand breaks. Nature 2004; 432:406-411.
22. Lisby M, Barlow JH, Burgess RC et al. Choreography of the DNA damage response: spatiotemporal relationships among checkpoint and repair proteins. Cell 2004; 118:699-713.
23. Stucki M, Clapperton JA, Mohammad D et al. MDC1 directly binds phosphorylated histone H2AX to regulate cellular responses to DNA double-strand breaks. Cell 2005; 123:1213-1226.
24. Fernandez-Capetillo O, Allis CD, Nussenzweig A. Phosphorylation of histone H2B at DNA double-strand breaks. J Exp Med 2004; 199:1671-1677.
25. Fernandez-Capetillo O, Celeste A, Nussenzweig A. Focusing on foci: H2AX and the recruitment of DNA-damage response factors. Cell Cycle 2003; 2:426-427.
26. Fernandez-Capetillo O, Lee A, Nussenzweig M et al. H2AX: the histone guardian of the genome. DNA Repair 2004; 3:959-967.
27. Lou Z, Minter-Dykhouse K, Franco S et al. MDC1 maintains genomic stability by participating in the amplification of ATM-dependent DNA damage signals. Mol Cell 2006; 21:187-200.
28. Franco S, Gostissa M, Zha S et al. H2AX prevents DNA breaks from progressing to chromosome breaks and translocations. Mol Cell 2006; 21:201-214.
29. Hirao A, Kong YY, Matsuoka S et al. DNA damage-induced activation of p53 by the checkpoint kinase Chk2. Science 2000; 287:1824-1827.
30. Matsuoka S, Rotman G, Ogawa A et al. Ataxia telangiectasia-mutated phosphorylates Chk2 in vivo and in vitro. Proc Natl Acad Sci USA 2000; 97:10389-10394.
31. Sorensen CS, Syljuasen RG, Falck J et al. Chk1 regulates the S-Phase checkpoint by coupling the physiological turnover and ionizing radiation-induced accelerated proteolysis of Cdc25A. Cancer Cell 2003; 3:247-258.
32. Zhao H, Watkins JL, Piwnica-Worms H. Disruption of the checkpoint kinase 1/cell division cycle 25A pathway abrogates ionizing radiation-induced S and G2 checkpoints. Proc Natl Acad Sci USA 2002; 99:14795-14800.
33. Falck J, Mailand N, Syljuasen RG et al. The ATM-Chk2-Cdc25A checkpoint pathway guards against radioresistant DNA synthesis. Nature 2001; 410:842-847.
34. Mailand N, Falck J, Lukas C et al. Rapid destruction of human Cdc25A in response to DNA damage. Science 2000; 288:1425-1429.
35. Saito S, Goodarzi AA, Higashimoto Y et al. ATM mediates phosphorylation at multiple p53 sites, including Ser(46), in response to ionizing radiation. J Biol Chem 2002; 277:12491-12494.
36. Meek DW. The p53 response to DNA damage. DNA Repair 2004; 3:1049-1056.
37. Lavin MF, Gueven N. The complexity of p53 stabilization and activation. Cell Death Differ 2006; 13:941-950.
38. Lavin MF, Kozlov S. DNA damage-induced signalling in ataxia telangiectasia and related syndromes. Radiother Oncol 2007; 83:231-237.
39. Stommel JM, Wahl GM. Accelerated MDM2 auto-degradation induced by DNA-damage kinases is required for p53 activation. EMBO J 2004; 23:1547-1556.
40. Pereg Y, Shkedy D, de Graaf P et al. Phosphorylation of Hdmx mediates its Hdm2- and ATM-dependent degradation in response to DNA damage. Proc Natl Acad Sci USA 2005; 102:5056-5061.
41. Chen L, Gilkes DM, Pan Y et al. ATM and Chk2-dependent phosphorylation of MDMX contribute to p53 activation after DNA damage. EMBO J 2005; 24:3411-3422.
42. Okamoto K, Kashima K, Pereg Y et al. DNA damage-induced phosphorylation of MdmX at serine 367 activates p53 by targeting MdmX for Mdm2-dependent degradation. Mol Cell Biol 2005; 25:9608-9620.
43. Dornan D, Shimizu H, Burch L et al. The proline repeat domain of p53 binds directly to the transcriptional coactivator p300 and allosterically controls DNA-dependent acetylation of p53. Mol Cell Biol 2003; 23:8846-8861.
44. Sherr CJ, Roberts JM. CDK inhibitors: positive and negative regulators of G1-phase progression. Genes Dev 1999; 13:1501-1512.
45. Lukas J, Lukas C, Bartek J. Mammalian cell cycle checkpoints: signalling pathways and their organization in space and time. DNA Repair 2004; 3:997-1007.
46. Hefferin ML, Tomkinson AE. Mechanism of DNA double-strand break repair by nonhomologous end joining. DNA Repair 2005; 4:639-648.

47. Weterings E, van Gent DC. The mechanism of nonhomologous end-joining: a synopsis of synapsis. DNA Repair 2004; 3:1425-1435.
48. Dynan WS, Yoo S. Interaction of Ku protein and DNA-dependent protein kinase catalytic subunit with nucleic acids. Nucleic Acids Res 1998; 26:1551-1559.
49. Hanakahi LA, Bartlet-Jones M, Chappell C et al. Binding of inositol phosphate to DNA-PK and stimulation of double-strand break repair. Cell 2000; 102:721-729.
50. Chen BP, Uematsu N, Kobayashi J et al. Ataxia telangiectasia mutated (ATM) is essential for DNA-PKcs phosphorylations at the Thr-2609 cluster upon DNA double strand break. J Biol Chem 2007; 282:6582-6587.
51. Goodarzi AA, Yu Y, Riballo E et al. DNA-PK autophosphorylation facilitates Artemis endonuclease activity. EMBO J 2006; 25:3880-3889.
52. Zhang X, Succi J, Feng Z et al. Artemis is a phosphorylation target of ATM and ATR and is involved in the G2/M DNA damage checkpoint response. Mol Cell Biol 2004; 24:9207-9220.
53. Zou L, Elledge SJ. Sensing DNA damage through ATRIP recognition of RPA-ssDNA complexes. Science 2003; 300:1542-1548.
54. Niida H, Nakanishi M. DNA damage checkpoints in mammals. Mutagenesis 2006; 21:3-9.
55. Wu X, Shell SM, Zou Y. Interaction and colocalization of Rad9/Rad1/Hus1 checkpoint complex with replication protein A in human cells. Oncogene 2005; 24:4728-4735.
56. Shiomi Y, Shinozaki A, Nakada D et al. Clamp and clamp loader structures of the human checkpoint protein complexes, Rad9-1-1 and Rad17-RFC. Genes Cells 2002; 7:861-868.
57. Kondo T, Wakayama T, Naiki T et al. Recruitment of Mec1 and Ddc1 checkpoint proteins to double-strand breaks through distinct mechanisms. Science 2001; 294:867-870.
58. Melo JA, Cohen J, Toczyski DP. Two checkpoint complexes are independently recruited to sites of DNA damage in vivo. Genes Dev 2001; 15:2809-2821.
59. Delacroix S, Wagner JM, Kobayashi M et al. The Rad9-Hus1-Rad1 (9-1-1) clamp activates checkpoint signaling via TopBP1. Genes Dev 2007; 21:1472-1477.
60. Lee J, Kumagai A, Dunphy WG. The Rad9-Hus1-Rad1 checkpoint clamp regulates interaction of TopBP1 with ATR. J Biol Chem 2007; 282:28036-28044.
61. Kumagai A, Lee J, Yoo HY et al. TopBP1 activates the ATR-ATRIP complex. Cell 2006; 124:943-955.
62. Ball HL, Myers JS, Cortez D. ATRIP binding to replication protein A-single-stranded DNA promotes ATR-ATRIP localization but is dispensable for Chk1 phosphorylation. Mol Biol Cell 2005; 16:2372-2381.
63. Chini CC, Chen J. Claspin, a regulator of Chk1 in DNA replication stress pathway. DNA Repair 2004; 3:1033-1037.
64. Cortez D, Guntuku S, Qin J et al. ATR and ATRIP: partners in checkpoint signaling. Science 2001; 294:1713-1716.
65. Wu X, Shell SM, Liu Y et al. ATR-dependent checkpoint modulates XPA nuclear import in response to UV irradiation. Oncogene 2007; 26:757-764.
66. Wu X, Shell SM, Yang Z et al. Phosphorylation of nucleotide excision repair factor xeroderma pigmentosum group A by ataxia telangiectasia mutated and Rad3-related-dependent checkpoint pathway promotes cell survival in response to UV irradiation. Cancer Res 2006; 66:2997-3005.
67. Batty DP, Wood RD. Damage recognition in nucleotide excision repair of DNA. Gene 2000; 241:193-204.
68. Cleaver JE, States JC. The DNA damage-recognition problem in human and other eukaryotic cells: the XPA damage binding protein. Biochem J 1997; 328 (Pt 1):1-12.
69. Adams KE, Medhurst AL, Dart DA et al. Recruitment of ATR to sites of ionising radiation-induced DNA damage requires ATM and components of the MRN protein complex. Oncogene 2006; 25:3894-3904.
70. Cuadrado M, Martinez-Pastor B, Murga M et al. ATM regulates ATR chromatin loading in response to DNA double-strand breaks. J Exp Med 2006; 203:297-303.
71. Jazayeri A, Falck J, Lukas C et al. ATM- and cell cycle-dependent regulation of ATR in response to DNA double-strand breaks. Nat Cell Biol 2006; 8:37-45.
72. Myers JS, Cortez D. Rapid activation of ATR by ionizing radiation requires ATM and Mre11. J Biol Chem 2006; 281:9346-9350.
73. Yoo HY, Kumagai A, Shevchenko A et al. Ataxia telangiectasia Mutated (ATM)-dependent Activation of ATR Occurs through Phosphorylation of TopBP1 by ATM. J Biol Chem 2007; 282:17501-17506.
74. Tomimatsu N, Tahimic CG, Otsuki A et al. Ku70/80 modulates ATM and ATR signaling pathways in response to DNA double strand breaks. J Biol Chem 2007; 282:10138-10145.
75. Stiff T, Walker SA, Cerosaletti K et al. ATR-dependent phosphorylation and activation of ATM in response to UV treatment or replication fork stalling. EMBO J 2006; 25:5775-5782.

76. Roos WP, Kaina B. DNA damage-induced cell death by apoptosis. Trends Mol Med 2006; 12:440-450.
77. Wu ZH, Shi Y, Tibbetts RS et al. Molecular linkage between the kinase ATM and NF-kappaB signaling in response to genotoxic stimuli. Science 2006; 311:1141-1146.
78. Shi Y, Venkataraman SL, Dodson GE et al. Direct regulation of CREB transcriptional activity by ATM in response to genotoxic stress. Proc Natl Acad Sci USA 2004; 101:5898-5903.
79. Dodson GE, Tibbetts RS. DNA replication stress-induced phosphorylation of cyclic AMP response element-binding protein mediated by ATM. J Biol Chem 2006; 281:1692-1697.
80. Kamer I, Sarig R, Zaltsman Y et al. Proapoptotic BID is an ATM effector in the DNA-damage response. Cell 2005; 122:593-603.
81. Zinkel SS, Hurov KE, Ong C et al. A role for proapoptotic BID in the DNA-damage response. Cell 2005; 122:579-591.
82. Kastan MB. Cell biology: A BID for the pathway. Nature 2005; 437:1103.
83. Wang X. The expanding role of mitochondria in apoptosis. Genes Dev 2001; 15:2922-2933.
84. Lane DP. Cancer. p53, guardian of the genome. Nature 1992; 358:15-16.
85. Urist M, Tanaka T, Poyurovsky MV et al. p73 induction after DNA damage is regulated by checkpoint kinases Chk1 and Chk2. Genes Dev 2004; 18:3041-3054.
86. Flinterman M, Guelen L, Ezzati-Nik S et al. E1A activates transcription of p73 and Noxa to induce apoptosis. J Biol Chem 2005; 280:5945-5959.
87. Flores ER, Tsai KY, Crowley D et al. p63 and p73 are required for p53-dependent apoptosis in response to DNA damage. Nature 2002; 416:560-564.
88. Rocco JW, Leong CO, Kuperwasser N et al. p63 mediates survival in squamous cell carcinoma by suppression of p73-dependent apoptosis. Cancer Cell 2006; 9:45-56.
89. Mancini M, Machamer CE, Roy S et al. Caspase-2 is localized at the Golgi complex and cleaves golgin-160 during apoptosis. J Cell Biol 2000; 149:603-612.
90. Zhivotovsky B, Samali A, Gahm A et al. Caspases: their intracellular localization and translocation during apoptosis. Cell Death Differ 1999; 6:644-651.
91. Lassus P, Opitz-Araya X, Lazebnik Y. Requirement for caspase-2 in stress-induced apoptosis before mitochondrial permeabilization. Science 2002; 297:1352-1354.
92. Bergeron L, Perez GI, Macdonald G et al. Defects in regulation of apoptosis in caspase-2-deficient mice. Genes Dev 1998; 12:1304-1314.
93. Robertson JD, Enoksson M, Suomela M et al. Caspase-2 acts upstream of mitochondria to promote cytochrome c release during etoposide-induced apoptosis. J Biol Chem 2002; 277:29803-29809.
94. Robertson JD, Gogvadze V, Kropotov A et al. Processed caspase-2 can induce mitochondria-mediated apoptosis independently of its enzymatic activity. EMBO Reports 2004; 5:643-648.
95. Kanu N, Behrens A. ATMIN defines an NBS1-independent pathway of ATM signalling. EMBO J 2007; 26:2933-2941.
96. McNees CJ, Conlan LA, Tenis N et al. ASCIZ regulates lesion-specific Rad51 focus formation and apoptosis after methylating DNA damage. EMBO J 2005; 24:2447-2457.
97. Matsuoka S, Ballif BA, Smogorzewska A et al. ATM and ATR Substrate Analysis Reveals Extensive Protein Networks Responsive to DNA Damage. Science 2007; 316:1160-1166.

Chapter 6

Chromosomal Instability in Ataxia Telangiectasia

Luitpold V. Distel* and Susann Neubauer

Abstract

Ataxia telangiectasia (AT) is a syndrome with multiple symptoms, of them one is the frequent and early onset of cancer. The disease is triggered by a mutation in the ATM-gene. The gene and its encoded protein, with 3056 amino acids, have different domains and functions. It plays a central role in the complex process of DNA double-strand break repair. The result of mis-repair and/or impaired DNA damage processing is due to chromosomal instability. An increased chromosomal instability is related to cancer incidence and normal tissue injury. Therefore chromosomal instability is an indicator of cancer risk. Here we examine the induced and spontaneous chromosomal aberrations studied by a three color FISH approach. Aberration rates, radiosensitivity and types of chromosomal aberrations in controls are compared with heterozygote and homozygote ATM carriers.

Introduction

The ATM gene is involved in several cellular functions that are triggered in response to spontaneous or induced DNA damage.[1] Chromosomal instability in Ataxia telangiectasia (AT) patients was first described by Syllaba and Henner[2] and Louis Barr.[3] Then AT was called Louis-Barr syndrome.

AT is an inherited autosomal recessive disease[4] and the ATM gene is located on 11q23.1. Cellular characteristics of AT are (i) the high number of spontaneous chromosomal breaks and rearrangements especially of the chromosomes 7 and 14 (5-10% of all cells analyzed showed this aberration) and (ii) high frequency of radiation induced chromosomal damage. Particularly most important breakpoints are 7p14, 7q35, 14q12 and 14q32, all of them are loci of the T-cell receptor gene.[5-8] The chromosomal instability in AT patients leads to a predisposition of these individuals to cancer.[9]

Chromosomal Instability and the G2-Assay

Studies on increased chromosomal instability after ionizing radiation in cultured cells was first carried out with fibroblasts in the so-called G2-assay.[10,11] Since fibroblasts require a skin biopsy and prolonged time for cell culture, this cytogenetic assay was later adapted to phytohemagglutinin (PHA) stimulated blood lymphocytes.[12,13] The G2-assay allowed the detection of chromatid breaks and gaps. An abnormal chromosomal response to ionizing radiation in the G2-phase of the cell cycle was demonstrated.[14-20] It is well known that cells from AT patients are defective in the cell cycle control, G1-S-Phase checkpoint, G2 phase checkpoint [9,21,22] and DNA-Repair.[23] All these impairments lead to chromosomal instability.

*Corresponding Author: Luitpold V. Distel—Department of Radiation Oncology, University Erlangen Nürnberg, Universitätsstr. 27, 91054 Erlangen, Germany.
Email: luitpold.distel@uk-erlangen.de

Molecular Mechanisms of Ataxia Telangiectasia, edited by Shamim I. Ahmad.

Chromosomal Instability and the G0-Assay

Another method to detect radiation induced chromosomal damage in blood lymphocytes is the G0/G1-assay.[24,25] Using this conventional cytogenetic assay, the chromosomal damage could be analyzed as dicentrics, ring chromosomes and open breaks. The FISH (fluorescence in situ hybridization) technique has provided a practical way to extent the scorable spectra of aberrations especially for translocations.[26,27] Chromosomal damage observed by one,[29] two or three color FISH[26,28-32] up to 24 color FISH[33,34] and M-FISH[35] approaches have been performed and the results available are fairly exciting and convincing.

Chromosomal Instability in ATM-Cells

In our own study several interesting questions arose concerning chromosomal instability in cells, heterozygous or homozygous, for ATM: (i) is there increased spontaneous chromosomal instability in heterozygous or homozygous ATM cells compared to healthy individuals? (ii) to what extent x-radiation enhances chromosomal instability in cells of AT heterozygotes and homozygotes? (iii) is there a broad inter-individual variability in AT patients? (iv) is there an altered spectrum of types of aberrations in the three groups examined? To answer these questions FISH assays were carried out.

By employing FISH assays analyses of chromosomal aberrations were carried out using whole chromosome painting of the chromosomes 1, 2 and 4. The three different painted chromosomes equalled 23% of the whole genome and where possible 1600 metaphases per individual were analyzed. The different types of aberrations were determined and summed up as breaks per metaphase. One break and isolated fragments equalled one break event; all reciprocal translocations, simple dicentrics and rings equaled two break events and complex aberrations equalled as many break events as theoretically necessary for the formation of the respective aberration. Complex aberrations were defined as exchange aberrations involving three or more break events in two or more chromosomes.

Frequency of Spontaneous Aberrations

Spontaneous chromosomal aberrations mean the occurrence of aberrations in unexposed cells. That means cells were unexposed either to ionizing radiation (IR) or to radiomimetic drugs. In healthy control cells aberrations are extremely rare events with about 0.01 ± 0.01 breaks per metaphase.[36-38]

Comparing the data for cells from 14 ATM heterozygous carriers and from 41 controls no distinction for spontaneous break events was found. Furthermore, there was no difference in the number of dicentrics, open breaks, translocations and complex aberrations or the fraction of aberrant cells. In comparison to these groups, in cells of ATM homozygotes, breaks per metaphase, open breaks, translocations and the number of aberrant cells were significantly increased, as well as a high number of dicentrics and complex aberrations were found (Figs. 1a and b). Furthermore, cells of ATM homozygotes showed a chromosomal damage three to six times higher compared to controls or heterozygotes. Open breaks appeared as the most outstanding aberrations within the spontaneous chromosomal damage. This type of aberration is hardly observed in control cells and was a rare event in cells obtained from ATM heterozygotes. Homozygotes showed a significant increase of open breaks being 30 times higher than in controls or heterozygotes.

In addition we investigated the range of the mean aberration rate (expressed as breaks per metaphase) among the individuals. For this the subjects were categorized according to the total number of breakpoints per metaphase and values were fitted using a Gaussian distribution. A broad inter-individual variability was observed in the group of controls and ATM heterozygotes (Fig. 1c and d). The highest variations and the highest number of all types of aberrations were detected in ATM homozygotes. Comparing the mean frequency of aberrations of control individuals and ATM homozygotes, the data increased by a factor of 2.8 (Fig. 1e). However, the level of aberrations was identical in cells from controls and cells of heterozygous ATM carriers. The frequency of spontaneous

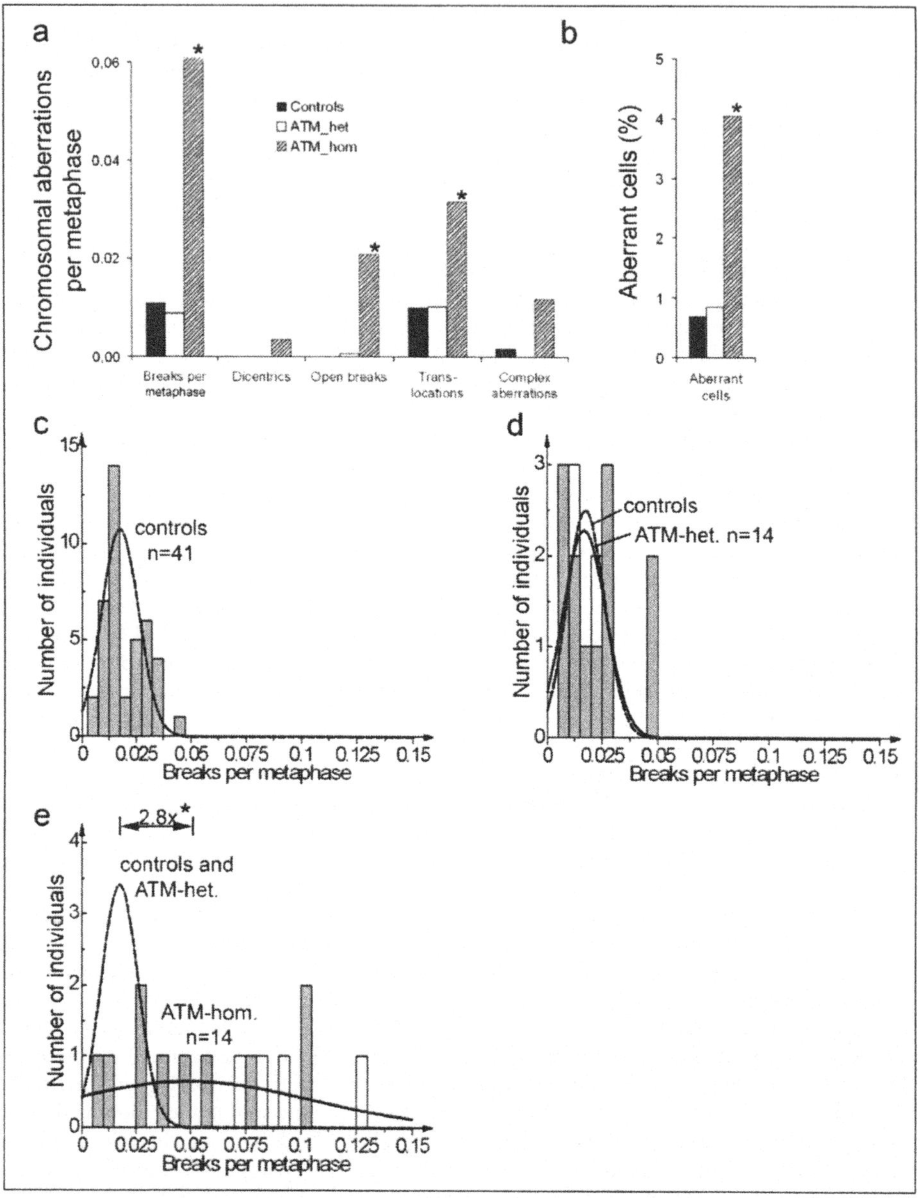

Figure 1. Spontaneous aberrations in cells derived from control individuals, individuals heterozygous and homozygous for ATM. Frequency of different types of spontaneous aberrations (a) and spontaneous rate of aberrant cells (b). Frequency distribution of spontaneous chromosomal aberrations as estimated in unexposed lymphocytes by three-color FISH. Chromosomal aberrations were scored as breaks per metaphase. Samples were derived from control individuals (c), individuals heterozygous for ATM (d) and homozygous for ATM (e) (open bars: peripheral blood lymphocytes and filled bars: lymphoblastoid cells). The data were fitted using a Gaussian distribution. Significant changes ($P < 0.05$) are marked with an asterisk. The arrows indicate the factor from mean to mean of the Gaussian distributions.

break events in some ATM homozygotes showed only a low level of aberrations being in the range of data from controls and heterozygotes. These results have been observed by others too.[35,39]

Frequency of Radiation Induced Aberrations

Because of the well documented sensitivity of AT patients to radiation and radiomimetic drugs, IR was used as the DNA damaging agent. The sparsely employed ionizing radiation distributed relatively small amounts of energy homogenously over the cell. Therefore DNA-damage was randomly distributed over the whole genome, followed by chromosomal aberrations. An advantage of ionizing radiation is that, after its passing through the cell, no toxic product remains except the DNA-damage. Hence it is possible to monitor exclusively the chromosomal instability as a consequence of DNA damage.

A moderate dose of 0.7Gy and other dose of 2Gy, equivalent to radiotherapy standard fractions, were employed to induce chromosomal aberrations. The higher dose leads to permanent cell cycle arrest in about 20-30% of the cells[40] and/or to apoptosis in about 1-4% of cells.[41] Chromosomal aberrations in cells from control, heterozygote and homozygote individuals increased with dose and the cytogenetic response was significantly higher in heterozygotes and homozygotes compared to controls (Fig. 2a). In the latter, in vitro irradiation with a dose of 2Gy led to an increased incidence of chromosomal aberrations by a factor of 22.6 compared to unirradiated cells. The aberration rate, expressed as breaks per metaphase, in cells from heterozygous and homozygous carriers increased by a factor of 54 and 40, respectively. These results reflect the high levels of chromosomal instability after genotoxic exposure among ATM carriers. ATM homozygotes have a highly increased chromosomal instability; however ATM heterozygotes exhibit a distinct increased instability under cytotoxic stress.

The inter-individual heterogeneity of radiation sensitivity is of great interest among healthy individuals and ATM heterozygous and homozygous carriers. In general population the variation may mainly be caused by polymorphisms and low penetrance mutations (which are relatively common but to the best of our knowledge have a low relative risk of the development of cancer). One of these low penetrance mutations in Western populations may be heterozygote ATM mutations with a prevalence of 0.5-1.0%.[42] Among individuals with mutations in the ATM gene the resulting chromosomal instability depends on a variety of reasons which are still not fully understood. The ATM gene spans ~150kBp of genomic sequence and contains 66 exons and has no frequent unique mutations and hot spots. The variability of chromosomal instability could be caused by the location of the mutation resulting in frameshift, nonsense, splicing or missense mutations in the gene. Most AT patients are compound heterozygote (it means the presence of two different mutant alleles at a particular gene locus, one on each chromosome of a pair) and these two mutations give a more complex situation to the level and activity of the ATM protein.

In spite of low levels of ATM mutated protein in cells from AT patients the remaining protein level and the residual kinase activity of the protein may have an impact on the variability of chromosomal instability. Consequently a significant variation in inter-individual sensitivity against all genotoxic agents would be expected. It was shown that individuals with an increased chromosomal instability in lymphocytes have an increased cancer risk. As well, cancer patients with high levels of chromosomal aberrations, after in vitro treatment with genotoxic agents, have an increased risk to develop normal tissue complications due to cancer treatment with ionizing radiation or chemotherapeutic agents.[43,44]

The high inter-individual variability is demonstrated after exposure of cells to 2 Gy ionizing radiation. In Figure 2b (left) are results from control individuals with a low chromosomal radiosensitivity to Figure 2b (right) are individuals with a higher radiosensitivity. This illustration fits well to a Gaussian distribution, with a mean of 0.4 ± 0.07 breaks per metaphase. Individuals outside the distribution exhibit a higher cytogenetic reaction.

However, in cells from ATM heterozygote carriers irradiation leads to a distinct 2.3 times higher breakage rate compared to controls (Fig. 2c). The cytogenetic instability shows a broad variation among heterozygotes with the lowest aberration rate being as low as aberration rates from controls

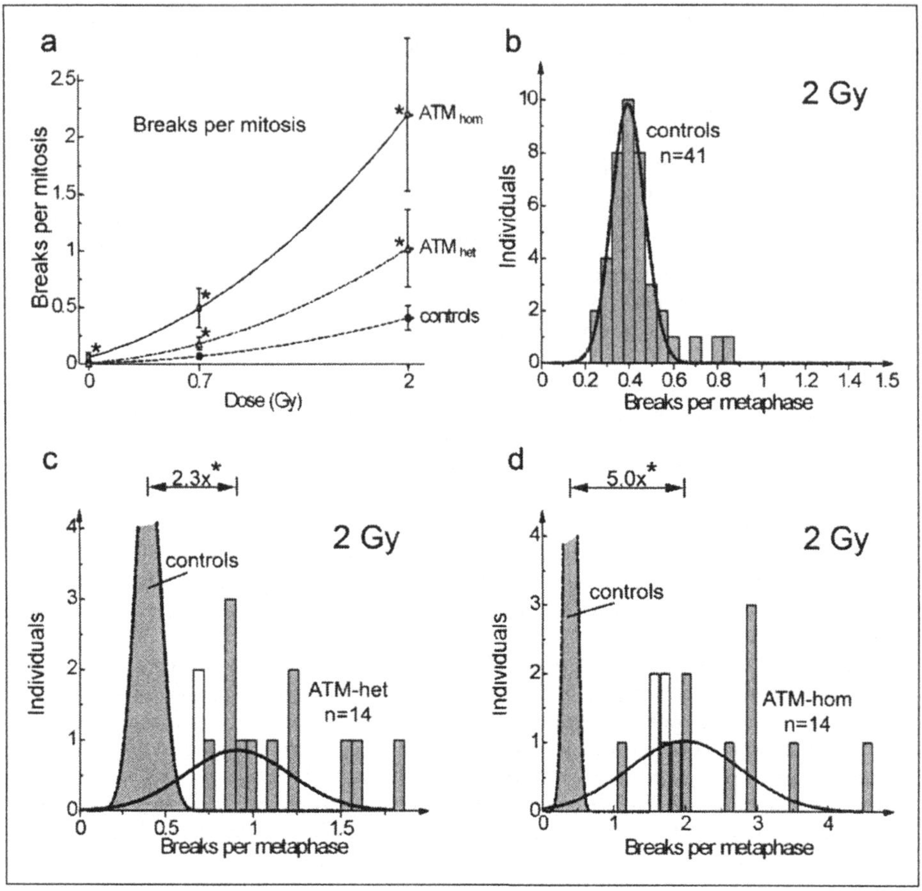

Figure 2. Ionizing radiation induced aberrations in cells derived from control individuals, individuals heterozygous and homozygous for ATM. Chromosomal aberrations were analyzed in in vitro irradiated cells by three-color FISH. Dose response-curve after 0.7Gy and 2.0Gy in vitro irradiation (a). Frequency distribution of the individual response of 2.0Gy ionizing radiation induced chromosomal aberrations estimated as breaks per metaphase. Samples were derived from control individuals (b), individuals heterozygous for ATM (c) and homozygous for ATM (d) (open bars: lymphocytes and filled bars: LCLs). The data were fitted using a Gaussian distribution. Significant changes ($P < 0.05$) are marked with an asterisk. The arrows indicate the factor from mean to mean of the Gaussian distributions.

and the highest value already as high as found in cells from homozygotes (1.99 ± 0.79). The mean value of the distribution of the heterozygotes is 2.3 and the homozygotes 5 times higher than the mean of the controls (Fig. 2d). Figure 3 summarizes the distribution of spontaneous and radiation induced chromosomal aberrations comparing the groups of control individuals, heterozygote and homozygote ATM individuals.

Types of Aberrations

There are several different types of aberrations induced by IR. It is of particular interest to find if different spectra of aberrations are induced in cells from ATM carriers compared to healthy controls.

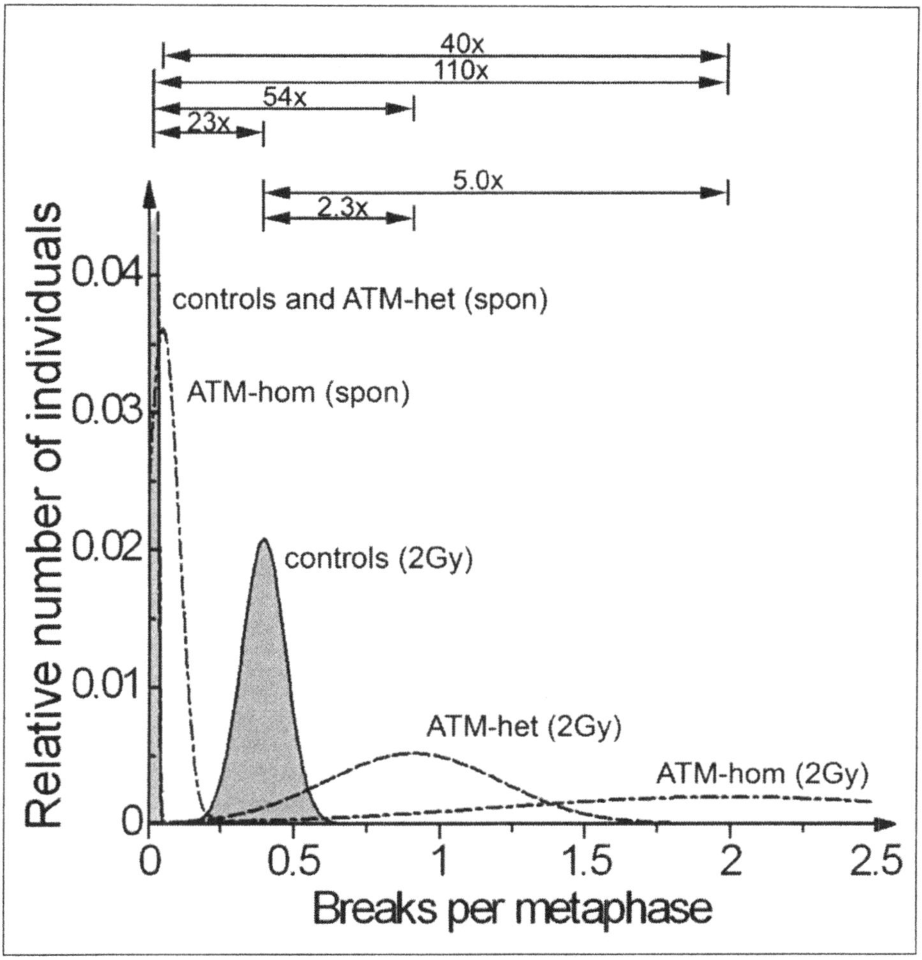

Figure 3. Spontaneous (spon) and radiation induced aberrations (2.0Gy) of all groups studied. Frequency distribution of chromosomal aberrations as analyzed in in vitro irradiated cells by three-color FISH. Chromosomal aberrations were scored as breaks per metaphase. Samples were derived from control individuals, individuals heterozygous and homozygous for ATM. The data were fitted using a Gaussian distribution. All data were combined from Figures 1 and 2. The arrows indicate the factor from mean to mean of the Gaussian distributions.

Performing a three-color FISH assay, different spectra of radiation induced chromosomal anomalies can be determined including unstable anomalies, e.g., dicentric and ring chromosomes with accompanying acentric fragments, a portion of complex rearrangements like dicentric chromosomes with an additional translocation, isolated fragments; also, open breaks as well as stable aberrations, such as reciprocal translocations, inversions and even complex anomalies like insertions can be seen. It is well-known that cell number decreases with the continuing number of cell divisions and chromosomal damage. However stable aberrations, independent from progressive numbers of cell cycles, are persistent in the cell till the cell dies.

In all three groups examined, healthy individuals, ATM heterozygous and homozygous individuals, a clear dose dependant response to in vitro irradiation of all types of chromosomal damage was observed. According to the chromosomal damage the number of aberrant cells increased with dose

(Fig. 4a-e). After in vitro irradiation all chromosomal aberrations in cells from individuals heterozygous and homozygous for ATM exceeded the chromosomal damage in cells from controls. The most frequent anomaly is the total number of translocations, i.e.,the total number of break events resulting from translocations (Fig. 4c). All observed aberrations, except dicentric chromosomes in cells from heterozygous and homozygous carriers increased significantly with dose compared to control individuals. With respect to the specific aberrations the most significant chromosomal anomaly in AT patients constitutes open breaks. In irradiated and unirradiated cells this anomaly reached the highest level of breaks compared to control individuals (Fig. 4b, g, h). Although open breaks were significantly increased in heterozygotes too, the most prominent increase was observed in cells from AT patients (Fig. 4b, f, g, h).

Complex aberrations were observed to be significantly increased in heterozygotes and in homozygotes. After in vitro irradiation the number of mitoses with complex rearrangements in healthy controls was very rare, varying between 0 and 37 per thousand metaphases. However in AT patients and especially in patients with a clinically defined radiation hypersensitivity this complex rearrangement highly increased, varying between 1.5 and 305 per thousand mitoses.[32] After a dose of 2Gy, the yield of aberrant cells detected with chromosomes 1, 2 and 4 in controls is about 20% while in AT patients about 70% of all cells analyzed have chromosomal damage. To adjust the chromosomal damage from the three chromosomes to a whole genome value, a factor of 2.9 is needed.[45] It is expected that chromosomal aberrations will be found in each cell from AT patients after a dose of 2Gy ionizing radiation. The high factors for open breaks and complex aberrations at 0.7Gy in homozygotes are caused by the very low value in controls and the high value in ATM homozygotes (Fig. 4g, h). This may cause an uncertainty for this factor, nevertheless there is a highly significant increase for these aberrations.

Summary

Without in vitro irradiation no difference in the cells from healthy control individuals and ATM heterozygotes is found concerning the frequency and the spectrum of chromosomal aberrations and the inter-individual variability. After in vitro irradiation, however, the ATM heterozygotes demonstrate distinct increased chromosomal instability compared to controls. In contrast, genotoxically stressed ATM homozygotes show an increased chromosomal instability with a distinct increase of complex aberrations and open breaks. The highest number of these aberrations was observed in irradiated cells.

Two major consequences are closely related to chromosomal instability. Firstly it is carcinogenesis after low levels of cytotoxic stress and, secondly, acute and chronic normal tissue injury displayed after intense cytotoxic stress, e.g., after therapy treatment. In the general population a correlation between chromosomal instability and cancer incidence exists.[43,46,47] Among cancer patients a link between chromosomal instability and acute and chronic tissue injury has been shown by several groups.[32,48-51] Extreme chromosomal instability in cells from homozygote ATM carriers may be closely related to increased cancer rates in about 30% of the AT patients.[52] A poor prognosis due to normal tissue injury after cancer therapy has been observed in several AT patients.[53-55] However it has not been proven yet that the chromosomal instability of AT homozygotes (ATM carriers) is directly correlated with cancer incidence or tissue injury. Confirmation of this correlation lacks mainly due to numbers of AT patients for cytogenetic testing.

In un-irradiated cells of heterozygote ATM carriers, no increased risk leading to chromosomal aberrations was found, yet under cytotoxic stress the frequency of aberrations is almost doubled. This increase appears at as low as 0.7Gy and presumably even at lower doses. It implicates that also ATM heterozygotes should be protected from cytotoxic stress so that they would not have an increased risk of cancer. However, there are a number of hidden sources for cytotoxic unavoidable stress; hence heterozygotes may have greater risk of developing cancer and therefore regular screening for early diagnosis of cancer may be carried out. A paradoxical situation is that in breast cancer screening, where regular mammograms with a dose of about 0.5mGy per breast are performed; this regular mammography may act as double edged sword by doubling the frequency of chromosomal aberrations and possibly an increased risk of cancer incidence.

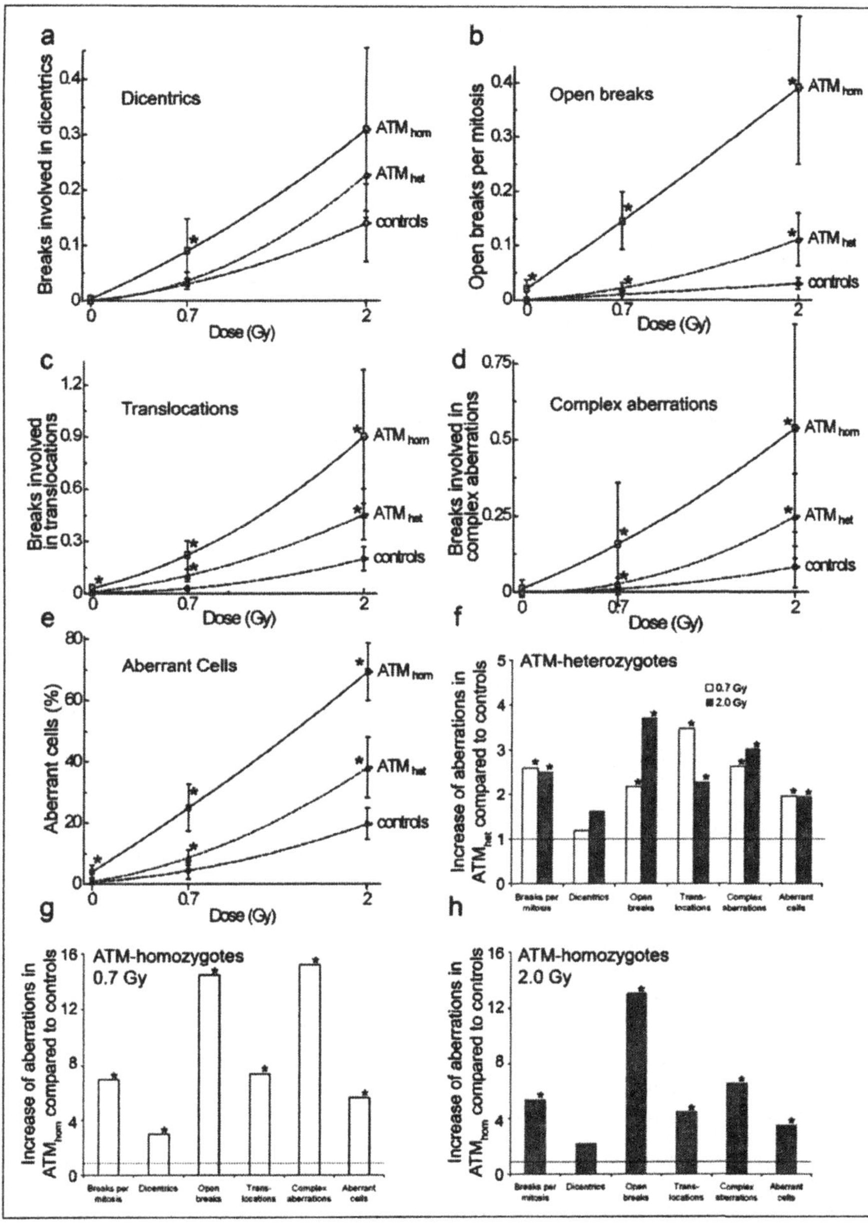

Figure 4. Dose response-curves for the induction of different types of aberrations in cells derived from control individuals, individuals heterozygous and homozygous for ATM after in vitro irradiation. Breaks involved in dicentrics (a), open breaks (b), breaks involved in translocations (c), breaks involved in complex aberrations (d) and aberrant cells by three-color FISH analysis as estimated in in vitro irradiated cells (e). The multiple of breakpoints involved in various types of aberrations and aberrant cells from heterozygous (f) and homozygous carriers of ATM (g, h) compared to control individuals after 0.7Gy (open bars) or 2.0Gy (striped bars) in vitro ionizing radiation is depicted. The horizontal line indicates the level of controls. Significant changes ($P < 0.05$) are marked with an asterisk.

In conclusion, AT-patients have an extreme increased chromosomal instability whether they are exposed to cytotoxic stress or not. The heterozygous ATM carriers, on the other hand, have under non cytotoxic conditions no increased risk for chromosomal instability but under cytotoxic stress they show a distinct increased chromosomal instability compared to controls.

References

1. Meyn MS. Ataxia telangiectasia and cellular responses to DNA damage. Cancer Res 1995; 55(24):5991-6001.
2. Syllaba L, Henner K. Contribution a l'independance de l'athetose double idiopathique et congenitale. Rev Neurol 1926; 1:541-562.
3. Louis-Bar D. Sur syndrome progressif comprenant des telangiectasies capillares cutanees et conjonctivales symetriques, a disposition naevoide et de troubles cerebelleux. Confin Neurol 1941; 4:32-42.
4. Gatti RA, Berkel I, Boder E et al. Localization of an ataxia telangiectasia gene to chromosome 11q22-23. Nature 1988; 336(6199):577-580.
5. Aurias A, Dutrillaux B, Buriot D et al. High frequencies of inversions and translocations of chromosomes 7 and 14 in ataxia telangiectasia. Mutat Res 1980; 69(2):369-374.
6. Aurias A, Dutrillaux B, Griscelli C. Tandem translocation t(14; 14) in isolated and clonal cells in ataxia telangiectasia are different. Hum Genet 1983; 63(4):320-322.
7. Humphreys MW, Nevin NC, Wooldridge MA. Cytogenetic investigations in a family with ataxia telangiectasia. Hum Genet 1989; 83(1):79-82.
8. O'Connor RD, Brown MG, Francke U. Immunologie and karyotypic studies in ataxia telangiectasia: specificity of break-points on chromosomes 7 and 14 in lymphocytes from patients and relatives. New York: Wiley, 1982; 259-270.
9. Rotman G, Shiloh Y. ATM: from gene to function. Hum Mol Genet 1998; 7(10):1555-1563.
10. Mitchell EL, Scott D. G2 chromosomal radiosensitivity in fibroblasts of ataxia telangiectasia heterozygotes and a Li-Fraumeni syndrome patient with radioresistant cells. Int J Radiat Biol 1997; 72(4):435-438.
11. Shiloh Y, Parshad R, Frydman M et al. G2 chromosomal radiosensitivity in families with ataxia telangiectasia. Hum Genet 1989; 84(1):15-18.
12. Chen P, Farrell A, Hobson K et al. Comparative study of radiation-induced G2 phase delay and chromatid damage in families with ataxia telangiectasia. Cancer Genet Cytogenet 1994; 76(1):43-46.
13. Sanford KK, Parshad R, Price FM et al. Enhanced chromatid damage in blood lymphocytes after G2 phase x irradiation, a marker of the ataxia telangiectasia gene. J Natl Cancer Inst 1990; 82(12):1050-1054.
14. Bender MA, Rary JM, Kale RP. G2 chromosomal radiosensitivity in ataxia telangiectasia lymphocytes. Mutat Res 1985; 152(1):39-47.
15. Bender MA, Rary JM, Kale RP. G0 chromosomal radiosensitivity in ataxia telangiectasia lymphocytes. Mutat Res 1985; 150(1-2):277-282.
16. Scott D, Barber JB, Levine EL et al. Radiation-induced micronucleus induction in lymphocytes identifies a high frequency of radiosensitive cases among breast cancer patients: a test for predisposition? Br J Cancer 1998; 77(4):614-620.
17. Scott D, Spreadborough AR, Roberts SA. Radiation-induced G2 delay and spontaneous chromosome aberrations in ataxia telangiectasia homozygotes and heterozygotes. Int J Radiat Biol 1994; 66(6 Suppl):S157-163.
18. Shiloh Y, Parshad R, Sanford KK et al. Carrier detection in ataxia telangiectasia. Lancet 1986; 1(8482):689-690.
19. Tchirkov A, Bay JO, Pernin D et al. Detection of heterozygous carriers of the ataxia telangiectasia (ATM) gene by G2 phase chromosomal radiosensitivity of peripheral blood lymphocytes. Hum Genet 1997; 101(3):312-316.
20. Wiencke JK, Wara DW, Little JB et al. Heterogeneity in the clastogenic response to X-rays in lymphocytes from ataxia telangiectasia heterozygotes and controls. Cancer Causes Control 1992; 3(3):237-245.
21. Lavin MF, Khanna KK. ATM: the protein encoded by the gene mutated in the radiosensitive syndrome ataxia telangiectasia. Int J Radiat Biol 1999; 75(10):1201-1214.
22. Deckbar D, Birraux J, Krempler A et al. Chromosome breakage after G2 checkpoint release. J Cell Biol 2007; 176(6):749-755.
23. Savitsky K, Bar-Shira A, Gilad S et al. A single ataxia telangiectasia gene with a product similar to PI-3 kinase. Science 1995; 268(5218):1749-1753.
24. Bauchinger M, Hug O. Chromosome aberrations in occupational radiation exposure in radiotherapy. Fortschr Geb Rontgenstr Nuklearmed 1968; 109(1):97-103.
25. Schmid E, Bauchinger M. Time relationships of radiation-induced chromosome aberrations in humans. Mutat Res 1969; 8(3):599-611.

26. Nakano M, Nakashima E, Pawel DJ et al. Frequency of reciprocal translocations and dicentrics induced in human blood lymphocytes by X-irradiation as determined by fluorescence in situ hybridization. Int J Radiat Biol 1993; 64(5):565-569.
27. Natarajan AT, Vyas RC, Darroudi F et al. Frequencies of X-ray-induced chromosome translocations in human peripheral lymphocytes as detected by in situ hybridization using chromosome-specific DNA libraries. Int J Radiat Biol 1992; 61(2):199-203.
28. Lucas JN, Awa A, Straume T et al. Rapid translocation frequency analysis in humans decades after exposure to ionizing radiation. Int J Radiat Biol 1992; 62(1):53-63.
29. Russell NS, Arlett CF, Bartelink H et al. Use of fluorescence in situ hybridization to determine the relationship between chromosome aberrations and cell survival in eight human fibroblast strains. Int J Radiat Biol 1995; 68(2):185-196.
30. Tucker JD, Lee DA, Ramsey MJ et al. On the frequency of chromosome exchanges in a control population measured by chromosome painting. Mutat Res 1994; 313(2-3):193-202.
31. Tucker JD, Senft JR. Analysis of naturally occurring and radiation-induced breakpoint locations in human chromosomes 1, 2 and 4. Radiat Res 1994; 140(1):31-36.
32. Neubauer S, Dunst J, Gebhart E. The impact of complex chromosomal rearrangements on the detection of radiosensitivity in cancer patients. Radiother Oncol 1997; 43(2):189-195.
33. Johannes C, Horstmann M, Durante M et al. Chromosome intrachanges and interchanges detected by multicolor banding in lymphocytes: searching for clastogen signatures in the human genome. Radiat Res 2004; 161(5):540-548.
34. Kuechler A, Neubauer S, Grabenbauer GG et al. Is 24-color FISH detection of in vitro radiation-induced chromosomal aberrations suited to determine individual intrinsic radiosensitivity? Strahlenther Onkol 2002; 178(4):209-215.
35. Martin M, Genesca A, Latre L et al. Radiation-induced chromosome breaks in ataxia telangiectasia cells remain open. Int J Radiat Biol 2003; 79(3):203-210.
36. Distel LV, Neubauer S, Keller U et al. Individual differences in chromosomal aberrations after in vitro irradiation of cells from healthy individuals, cancer and cancer susceptibility syndrome patients. Radiother Oncol 2006; 81(3):257-263.
37. Neubauer S, Arutyunyan R, Stumm M et al. Radiosensitivity of ataxia telangiectasia and Nijmegen breakage syndrome homozygotes and heterozygotes as determined by three-color FISH chromosome painting. Radiat Res 2002; 157(3):312-321.
38. Verdorfer I, Neubauer S, Letzel S et al. Chromosome painting for cytogenetic monitoring of occupationally exposed and non exposed groups of human individuals. Mutat Res 2001; 491(1-2):97-109.
39. Stumm M, Neubauer S, Keindorff S et al. High frequency of spontaneous translocations revealed by FISH in cells from patients with the cancer-prone syndromes ataxia telangiectasia and Nijmegen breakage syndrome. Cytogenet Cell Genet 2001; 92(3-4):186-191.
40. Perez A, Grabenbauer GG, Sprung CN et al. Potential for the G2/M arrest assay to predict patient susceptibility to severe reactions following radiotherapy. Strahlenther Onkol 2007; 183(2):99-106.
41. Wistop A, Keller U, Sprung CN et al. Individual radiosensitivity does not correlate with radiation-induced apoptosis in lymphoblastoid cell lines or CD3+ lymphocytes. Strahlenther Onkol 2005; 181(5):326-335.
42. Renwick A, Thompson D, Seal S et al. ATM mutations that cause ataxia telangiectasia are breast cancer susceptibility alleles. Nat Genet 2006; 38(8):873-875.
43. Hagmar L, Brogger A, Hansteen IL et al. Cancer risk in humans predicted by increased levels of chromosomal aberrations in lymphocytes: Nordic study group on the health risk of chromosome damage. Cancer Res 1994; 54(11):2919-2922.
44. Hagmar L, Stromberg U, Bonassi S et al. Impact of types of lymphocyte chromosomal aberrations on human cancer risk: results from Nordic and Italian cohorts. Cancer Res 2004; 64(6):2258-2263.
45. Gebhart E, Neubauer S, Schmitt G et al. Use of a three-color chromosome in situ suppression technique for the detection of past radiation exposure. Radiat Res 1996; 145(1):47-52.
46. Hagmar L, Bonassi S, Stromberg U et al. Chromosomal aberrations in lymphocytes predict human cancer: a report from the European Study Group on Cytogenetic Biomarkers and Health (ESCH). Cancer Res 1998; 58(18):4117-4121.
47. Bonassi S, Znaor A, Norppa H et al. Chromosomal aberrations and risk of cancer in humans: An epidemiologic perspective. Cytogenet Genome Res 2004; 104(1-4):376-382.
48. Hoeller U, Borgmann K, Bonacker M et al. Individual radiosensitivity measured with lymphocytes may be used to predict the risk of fibrosis after radiotherapy for breast cancer. Radiother Oncol 2003; 69(2):137-144.
49. Keller U, Grabenbauer G, Kuechler A et al. Cytogenetic instability in young patients with multiple primary cancers. Cancer Genet Cytogenet 2005; 157(1):25-32.

50. Lavin MF, Bennett I, Ramsay J et al. Identification of a potentially radiosensitive subgroup among patients with breast cancer. J Natl Cancer Inst 1994; 86(21):1627-1634.
51. Scott D. Chromosomal radiosensitivity and low penetrance predisposition to cancer. Cytogenet Genome Res 2004; 104(1-4):365-370.
52. Chun HH, Gatti RA. Ataxia telangiectasia, an evolving phenotype. DNA Repair (Amst) 2004; 3(8-9):1187-1196.
53. Abadir R, Hakami N. Ataxia telangiectasia with cancer. An indication for reduced radiotherapy and chemotherapy doses. Br J Radiol 1983; 56(665):343-345.
54. Hart RM, Kimler BF, Evans RG et al. Radiotherapeutic management of medulloblastoma in a pediatric patient with ataxia telangiectasia. Int J Radiat Oncol Biol Phys 1987; 13(8):1237-1240.
55. Pritchard J, Sandland MR, Breatnach FB et al. The effects of radiation therapy for Hodgkin's disease in a child with ataxia telangiectasia: A clinical, biological and pathologic study. Cancer 1982; 50(5):877-886.

CHAPTER 7

Cell Cycle Defects and Apoptosis in Ataxia Telangiectasia

Deborah Wilsker* and Fred Bunz

Abstract

ATM is a key regulator of both cell cycle checkpoints and apoptosis. Cells lacking ATM are defective in critical responses to damaged DNA and particularly to the double strand DNA breaks caused by ionizing radiation (IR). Depending on the cell type, ATM-deficient cells are extremely sensitive to IR or, alternatively, resistant to IR-induced apoptosis. The many different consequences of ATM activation that have been observed reveal the numerous pathways that are controlled by ATM.

Introduction

AT is characterized by a high degree of pleiotropy and accordingly ATM activation has numerous cellular consequences. ATM has established roles in diverse cellular pathways, including those that modulate cell cycle regulation, DNA repair and apoptosis. The roles of ATM in DNA damage recognition and DNA repair are discussed at length in other chapters. We focus here on the pathways that effect ATM-dependent cell cycle regulation and apoptosis.

DNA Damage and Cell Cycle Progression

ATM is a central component of the cellular response to double strand DNA breaks (DSBs). DSBs represent a potent challenge to both genomic integrity and cellular survival. DSBs can be caused by exogenous sources such as ionizing radiation (IR) and radiomimetic chemicals (collectively known as clastogens) or endogenous physiological processes such as replication and recombination. Recognition of DSBs by the cell triggers numerous responses that result in the repair of DSBs or apoptosis. DSBs that remain unrepaired or are repaired inaccurately can give rise to de novo mutations that promote tumorigenesis.

Successful repair of a DSB requires a concerted halt in the cell division cycle and the activation of DNA repair pathways. A useful analogy is a car with a damaged engine. In such a case, the engine must be shut down before repairs can be safely attempted. Similarly, cell processes such as DNA replication and mitosis must be arrested before chromosomes can be repaired. A genetically defined pathway by which cells halt their cell cycle in the presence of DNA damage is known as a "checkpoint". Checkpoints inactivate the enzymes that promote cell cycle transitions and function to prevent the replication and segregation of damaged chromosomes.

Cell cycle phases are characterized by the enzymatic activity of Cyclin-Dependent Kinases (CDKs). CDKs bind cyclin protein subunits that confer activity during defined phases of the

*Corresponding Author: Deborah Wilsker—The Department of Radiation Oncology and Molecular Radiation Sciences and the Sidney Kimmel Comprehensive Cancer Center, Johns Hopkins University School of Medicine, Baltimore, Maryland, USA.
Email: dwilske1@jhmi.edu

Molecular Mechanisms of Ataxia Telangiectasia, edited by Shamim I. Ahmad.

cell cycle.[1] The sequential activation of CDKs by cyclins ensures ordered progression through the cell cycle.

CDKs are regulated in at least three ways. First, CDKs are activated by cyclin binding. Unbound CDK molecules are enzymatically inactive. Second, CDKs are regulated by the binding of Cyclin-Dependent Kinase Inhibitors (CDKIs). Third, CDKs contain highly conserved residues that regulate kinase activity through phosphorylation and dephosphorylation. Wee1 and MYT1 kinases phosphorylate and inhibit CDK activity while Cell division cycle 25 (Cdc25) phosphatases can activate CDK activity by removing the inhibitory phosphate.

Of the three Cdc25 phosphatases, Cdc25A, B and C, Cdc25A appears to play the most prominent role in cell cycle regulation after DNA damage.[2] Upon DNA damage, Cdc25A is phosphorylated and degraded by ubiquitin-mediated proteolysis.[3] The inactivation of Cdc25A results in the persistence of inhibitory phosphates on CDKs, thereby blocking cell cycle transitions. Cdc25B and C are also regulated by phosphorylation which results in their exclusion from the nucleus. Exclusion from the nucleus, in turn, prevents Cdc25B and C from accessing and activating their CDK substrates, leading to a halt in the cell cycle.

The cell cycle features the replication of the genome and division of a cell into two daughter cells. DNA replication and cell division occur in discrete phases. Chromosomes are replicated during S (synthesis) phase. During M (mitosis) phase, the orderly segregation of chromosomes into daughter cells occurs. S- and M-phases are separated by two gap phases, G_1 and G_2. G_1 precedes DNA replication and G_2 directly follows the completion of DNA replication and precedes mitosis. It is during these gap phases that cells are particularly sensitive to antiproliferative signals.

The three best understood cell cycle checkpoints occur at the transition between G_1/S, within S-phase and at the G_2/M transition. Of these three, the intra-S checkpoint is somewhat different because it is invariably transient while the other two can be very stable. The G_1/S checkpoint prevents damaged chromosomes from being replicated while the G_2/M checkpoint prevents segregation of damaged chromosomes in mitosis.[4,5]

ATM Activates Checkpoints

Like all pathways, DNA damage checkpoint pathways have three main components: sensors, transducers and effectors. Sensor proteins in the DNA damage signaling network belong to a protein family known as the phosphoinositide 3-kinase-like kinases (PIKKs). These proteins share sequence homology to the catalytic domain of the PI3 kinase family but have protein kinase activity rather than the lipid kinase activity exhibited by typical PI3 kinases. PIKK subfamily members are serine/threonine protein kinases. Both ATM and the related PIKK ataxia telangiectasia mutated and Rad3 related (ATR) protein have central roles in sensing of DNA damage and in the activation of downstream transducers.

ATM is the primary responder to DSBs. Recent studies have shed a great deal of light on the molecular details of ATM activation (see Chapter 7). ATM is a 300 kDa protein kinase that, in the absence of DNA damage, exists primarily in oligomeric form. Following exposure to IR, ATM is converted from an inactive oligomer to an active monomer, a conformational change that is accompanied by autophosphorylation.[6] In contrast to ATM, ATR primarily responds to agents that cause the stalling of DNA replication forks, such as ultraviolet light and drugs that inhibit nucleotide metabolism.

Recent findings indicate a surprising degree of interplay between ATM and ATR.[7] ATM promotes the recruitment of ATR to DSB sites and thereby facilitates its activation. Conversely, it appears that ATM can be activated in an ATR-dependent manner by DNA synthesis inhibition.[8,9] While it was formerly thought that ATM and ATR control entirely distinct, parallel pathways, it has recently become apparent that ATM and ATR work together to coordinate downstream responses to diverse forms of DNA damage.

Phosphorylation of many ATM substrates requires more than activation of the ATM monomer. Activated ATM must localize to the DSB sites where several of its critical substrates accumulate. One such substrate is Nbs1, a component of the Mre11/Rad50/Nbs1 (MRN) complex. This complex assembles at the DSB site independently of ATM[6] and is required for localization of ATM kinase to sites of DSBs. In cells lacking Nbs1, the ATM monomer can be activated and can phosphorylate its nucleoplasmic substrates such as p53, but cannot localize to sites of DNA damage or phosphorylate its critical substrates there.[10,11] Mutations in Nbs1 or Mre11 cause the human diseases Nijmegen Breakage Syndrome and Ataxia Telangiectasia-like Disorder (ATLD), respectively. These diseases exhibit significant clinical overlap with AT, suggesting that ATM and the MRN complex function in a common pathway.

Once activated, ATM phosphorylates over 700 proteins on more than 900 sites.[12] Among these many substrates are proteins that are known to function in cell cycle control, DNA repair and apoptosis. Two well characterized ATM and ATR target proteins are the checkpoint kinases, Chk1 and Chk2. These structurally unrelated proteins function as the primary transducers of PIKK signals. Both proteins are phosphorylated by ATM in response to DSBs, but Chk2 activation appears to be ATM-dependent.[13,14] Chk1 is primarily phosphorylated by ATR in response to replication stress. Upon activation by ATM and ATR, Chk2 and Chk1 phosphorylate downstream substrates, including, critically, the Cdc25 phosphatase family. Thus, via the transducers Chk1 and Chk2, the signal triggered by a DNA break is converted from a local one to one that reaches distant proteins in the cell.[15]

One ATM substrate that is critical to the checkpoint response is SMC1 (structural maintenance of chromosomes 1), a subunit of the cohesin protein complex involved in sister chromatid cohesion. In response to DSBs, ATM directly phosphorylates SMC1 on Ser957 and Ser966 at the DSB site.[10,16] Human and mouse cells, expressing mutant SMC1 (containing nonphosphorylatable amino acid substitutions at positions 957 and 966) exhibit an intra-S-phase checkpoint defect, hypersensitivity to IR and increased number of chromosome gaps and breaks. These phenotypes are similar to those caused by ATM-deficiency.[6,10,11]

Phosphorylation of SMC1 by ATM requires Nbs1 and BRCA1 (Breast Cancer 1, early onset).[10,11,16] IR-induced phosphorylation of SMC1 does not occur in cells lacking full-length Nbs1 or BRCA1 despite the fact that ATM is activated in these cells.[10] This finding prompted further studies of ATM subcellular localization which revealed that BRCA1 and Nbs1 are required for localization of activated ATM to sites of DNA strand breaks.[11]

ATM phosphorylates BRCA1 at several sites. BRCA1 interacts with an extensive list of proteins that are critical tumor suppressors, oncogenes, DNA damage repair and cell cycle regulators. Interestingly, the particular site upon which BRCA1 is phosphorylated by ATM dictates whether it will act as a regulator of the intra-S-Phase checkpoint or the G2-M checkpoint.[17] Through its interactions with numerous proteins downstream of ATM and ATR, BRCA1 is an important regulator of cell cycle progression.[18] Additionally during the intra-S-Phase checkpoint FANCD2 (Fanconi anemia complementation group D2) is phosphorylated in an ATM-dependent manner leading to its mono-ubiquitination and colocalization with BRCA1 to DSB sites where it may interact with other DNA damage response proteins involved in recognition and repair.[19,20] FANCD2 is a member of a multi-protein complex, defects of which lead to the cancer susceptibility disorder called Fanconi anemia.[21]

ATM and the G_1/S Checkpoint

The G_1/S checkpoint prevents the replication of damaged chromosomes by preventing the $G_1 \rightarrow$ S transition. The mechanism by which proliferating cells enter S-phase has been well-established. Briefly, the $G_1 \rightarrow$ S transition is characterized by the activation of Cdk4 and then Cdk2 in late G_1 and early S-phase. A key substrate of activated Cdk4 and Cdk2 is Rb, the product of the retinoblastoma gene. Rb physically associates with and blocks the activity of E2F, a transcription factor that induces expression of several genes required for S-phase progression. At the transition from G_1 to S, pRb is successively hyperphosphorylated and dissociates from E2F, which is then free to transcriptionally upregulate genes that promote the onset of DNA replication. ATM sets into play two major pathways that are activated in parallel and lead to initiation and maintenance of the G_1/S checkpoint.[3,22] These pathways lead to (1) the degradation of dual specificity protein phosphatase Cdc25A and (2) the accumulation of the CDKI p21.

Initiation of G_1/S Arrest

DSBs trigger the rapid degradation of the phosphatase Cdc25A.[13,23,24] Cdc25A is phosphorylated in an ATM and Chk2-dependent manner and this phosphorylation leads to its degradation.[25] Interestingly, the phosphorylation and degradation of Cdc25A plays a role in the progression of the unperturbed cell cycle as well.[26] Although the mechanisms of Cdc25A phosphorylation following IR remain incompletely understood, its DNA damage-dependent degradation results in CDK inactivation and results in cell cycle arrest.

Maintenance of G_1/S Arrest

While the degradation of Cdc25A triggers IR-induced G_1/S arrest, maintenance of this checkpoint is dependent on a pathway mediated by the transcriptional transactivator p53. The connection between ATM and p53 was noted prior to the identification and cloning of the ATM gene when it was observed that p53 stabilization after IR is defective in cells derived from AT patients.[27] Subsequently it was shown that, in response to DSBs, ATM phosphorylates p53 directly on Ser15 and contributes to stabilization of the protein.[28] ATR, Chk1 and Chk2 have also been shown to phosphorylate p53;[29-32] the relative importance of these kinase/substrate relationships remain unclear.

In response to its stabilization, p53 transcriptionally transactivates many downstream target genes including the CDK inhibitor p21. p21 binds to and inhibits the S-phase promoting Cdk2/cyclin E complex and the Cdk4/Cyclin D complex, preventing the phosphorylation of pRb. Thus, pRb/E2F remains in its active, growth suppressive state and prevents progression through G_1. Since ATM is required for optimal phosphorylation of p53 after IR, the G_1/S checkpoint is affected in AT cells.[27,33] This ATM-p53-p21 pathway can take several hours to be activated but is sustained for a long time as compared with the relatively transient response of Cdc25A degradation and is required for the maintenance of G_1/S arrest.[3,5,34,35](Fig. 1). In addition to being able to sustain G_1/S arrest for longer durations, the ATM-p53 signaling pathway is involved in inducing apoptosis (see below).

The Intra-S Checkpoint

The intra-S checkpoint is activated if DNA damage occurs during DNA replication and results in a transient halt in DNA replication. The molecular mechanism of this checkpoint has recently been elucidated. DSBs lead to rapid degradation of Cdc25A protein and to the persistence of the inactive phosphorylated form of Cdk2. Inactive CDK2 is incapable of phosphorylating its downstream target Cdc45, a crucial factor for the initiation of eukaryotic DNA replication. Cdc45 is essential for the loading of DNA polymerase α at replication origins and so its inhibition by the intra-S checkpoint response prevents the firing of late-firing origins.

As one might expect, Nbs1 is also a necessary component of the intra-S checkpoint. In particular Nbs1 is important in the recruitment of ATM to DSBs.[36,37] Once localized to a DSB, ATM phosphorylates Nbs1 on two serine residues at positions 278 and 343 after IR.[38-40] Mutations at these sites to alanines, that can not be phosphorylated, exhibit defective S-phase checkpoint in response to IR and mutant Nbs1 fails to rescue the S-phase checkpoint in NBS cells.[38,40] These results indicate that Nbs1 phosphorylation by ATM is required for the S-phase checkpoint in response to IR. In addition to its role in localizing ATM to DSB sites through the MRN complex, Nbs1 may act as a mediator in facilitating the phosphorylation of other ATM downstream substrates.

The loss of the intra-S checkpoint by ATM mutation causes a cellular phenotype known as Radioresistant DNA Synthesis (RDS). In response to IR, DNA replication rapidly comes to a halt in cells with an intact S-phase checkpoint. Many years ago, it was observed that fibroblasts from AT patients fail to normally halt DNA synthesis in response to IR, despite extensive DNA damage.[41]

AT patients are extremely sensitive to IR and AT fibroblasts are hypersensitive to killing by ionizing radiation.[42] ATM defects are most notable in S-phase rather than other checkpoints. Other proteins are also known to have RDS phenotypes when depleted or absent; these include BRCA1, Nbs1 and FANCD2. However patients with mutations in these genes are only slightly sensitive to the effects of IR whereas AT patients are extremely sensitive.

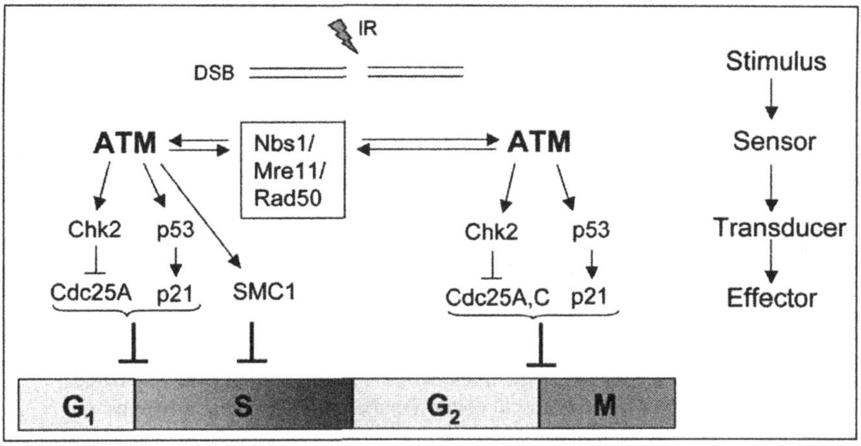

Figure 1. ATM-dependent DNA damage checkpoint responses. A schematic of some checkpoint signaling events discussed in this chapter and the cell cycle phase in which they are activated.

G_2/M Checkpoint

The G_2/M checkpoint prevents the segregation of damaged chromosomes in mitosis. Defects in this checkpoint are thought to contribute to the loss of genomic stability.[5]

ATM is an important modulator of the G_2/M checkpoint. AT cells exhibit defective G_2/M arrest after IR.[43] Direct analysis of radiation-induced DNA damage, before and after the G_2/M transition, revealed that while AT cells may incur the same number of DSBs in G_2, they have increased damage of metaphase chromosomes, as compared to normal cells.[44] This result suggests that the ATM-mediated G_2/M checkpoint may function to allow time for repair of chromosomal breaks. If DSBs are present during G_2, ATM activation leads to inhibition of Cdc25 activity and subsequently the inactivation of the critical target of the G_2/M checkpoint: CDK1/cyclin B.[5]

Inhibition of Cdc25 not only initiates G2/M arrest but the maintenance phase of the G_2/M checkpoint may rely on activation of p53 and the subsequent upregulation of cell cycle inhibitors. The ATM-p53-p21 signaling pathway is required to sustain G_2/M arrest after IR.[45] As in the case of the G_1/S checkpoint, the CDK inhibitor p21 also plays a role in sustaining G_2/M arrest after IR. Stabilization of p53 protein is critical for stable arrest of both of these checkpoints.

Recent studies have identified a novel role for Che-1, a RNA polymerase II binding protein, in the ATM-mediated stabilization of p53 and the maintenance of the G_2/M checkpoint.[46] Che-1 regulates p53 by controlling its transcription. Che-1 is a substrate of ATM and Chk2 and its stability is severely impaired in fibroblasts from AT patients when treated with doxorubicin, a radiomimetic drug that induces double strand breaks. Therefore, analysis of Che-1 reveals yet another component of the ATM -p53 regulatory response to DSBs.

ATM phosphorylation of BRCA1 can lead to activation of the G_2/M checkpoint in addition to activation of the intra-S checkpoint,[17] as discussed above. BRCA1 is important in several aspects of the G_2/M transition such as the recruitment of ATM to sites of DSB, the activation of downstream target proteins and it also regulates transcriptional activation of cell cycle inhibitors such as Gadd45.[47]

In addition to the Cdc25 and p53 responses, recent findings indicate a potential third cell cycle checkpoint pathway mediated by ATM and ATR. The p38MAPK/MK2 pathway is a general stress response pathway that responds to cellular stimuli such as cytokines and hyperosmolarity. This greatly studied pathway has recently been discovered to play a role in the G_2/M checkpoint and to be activated in response to drugs that cause DNA damage including those that cause DSBs. In the absence of p53, ATM and ATR may alternatively signal via the p38MAPK/MK2 stress response pathway in response to DSB at the G_2/M transition.[48] These results suggest that ATM-dependent signaling events may enhance the survival of p53-null cancer cells.

DNA Damage Induced Cell Death by ATM-Mediated Apoptosis

The preceding sections describe one potential outcome of ATM activation: cell cycle arrest. A second potential outcome is death. Apoptosis is the genetically programmed death and breakdown of a cell in response to defined stimuli.[49] Apoptosis is required for the maintenance of tissue homeostasis. ATM plays a key role in this process.

The most obvious link between ATM and apoptosis is p53. The amount of DSBs may affect the extent to which ATM phosphorylates p53. The level at which p53 is activated in response to DNA damage appears to affect whether p53 induces transcription of genes that cause cell cycle arrest such as p21, or induces pro-apoptotic genes such as BAX (BCL2-associated X protein), PUMA (p53 upregulated modulator of apoptosis) and FAS receptor.[50,51] Whether p53 activates apoptosis or cell cycle arrest depends on several key cellular factors that have only begun to be elucidated.[52]

The downstream substrates that are phosphorylated by ATM depend on the type of DNA damage, the extent of this damage and the particular type of cell incurring the damage.[53] It has become clear that the cellular context can be crucial to determining the biological outcome. For example, fibroblasts derived from people or mice that are ATM-null are hypersensitive to the lethal effects of IR, however ATM-deficient cells from the developing nervous system are resistant to it.[54,55]

One example of the tissue specific role of ATM in apoptosis has recently been examined in lymphoid cells. AT patients suffer from cancer predisposition, especially in cells of the immune system and frequently develop lymphoma and leukemia. The sensing of DSBs is critical to immune system development due to the DSBs generated during VDJ recombination. ATM clearly plays a central role sensing DSBs during development and in eliminating lymphoid clones via apoptosis. Recent studies have revealed that ATM deficiency causes upregulation of FLIP, a well-known inhibitor of Fas-induced apoptosis.[56] Resistance of AT lymphoid cells to Fas-induced apoptosis suggests a mechanism by which ATM loss alters tissue homeostasis. This mechanism of apoptotic regulation suggests a particularly specific role for ATM in lymphoid cells that is independent of exogenous DNA damage and may account in part for the AT pathology in the immune system.

ATM-Mediated Apoptosis in Neuronal Cells

Neurological defects are a cardinal feature of AT and the role of ATM in the cells of the central nervous system is therefore of particular interest. Whether DNA damage induces cell cycle arrest or apoptosis in neuronal cells is influenced by the developmental stage of the nervous system. During neural development when there is a prolonged period of cell division, DNA damage predominantly activates apoptosis instead of DNA repair. However, in the mature, differentiated nervous system cell cycle arrest, followed by DNA repair, is more likely to occur.[57] In fact, ATM is highly expressed in the developing nervous system but is at low levels in the adult CNS.[58] Importantly, it appears that a crucial function of ATM in the developing nervous system is to eliminate neural cells that incur excessive DNA damage[54] (Fig. 2). A failure to do so results in neural cells with compromised genetic integrity and cellular function. ATM-null neural cells are not eliminated during neural development and over time may continue to acquire genetic lesions that lead to compromised cellular function and neurodegeneration associated with AT.[59] Thus, the resistance of ATM-deficient neuronal cells to DNA damage-induced apoptosis could be the underlying biological basis for neurodegeneration in AT.[59]

Conclusion

ATM is a central mediator of cellular responses to DNA damage. When double-strand breaks are introduced ATM is rapidly activated and is free to circulate in the cell and activate substrates such as p53 in the nucleoplasm. Nbs1 and BRCA1 proteins attract activated ATM to the localized sites of DNA breaks where it can phosphorylate substrates such as SMC1 that are critical in minimizing chromosomal aberrations.

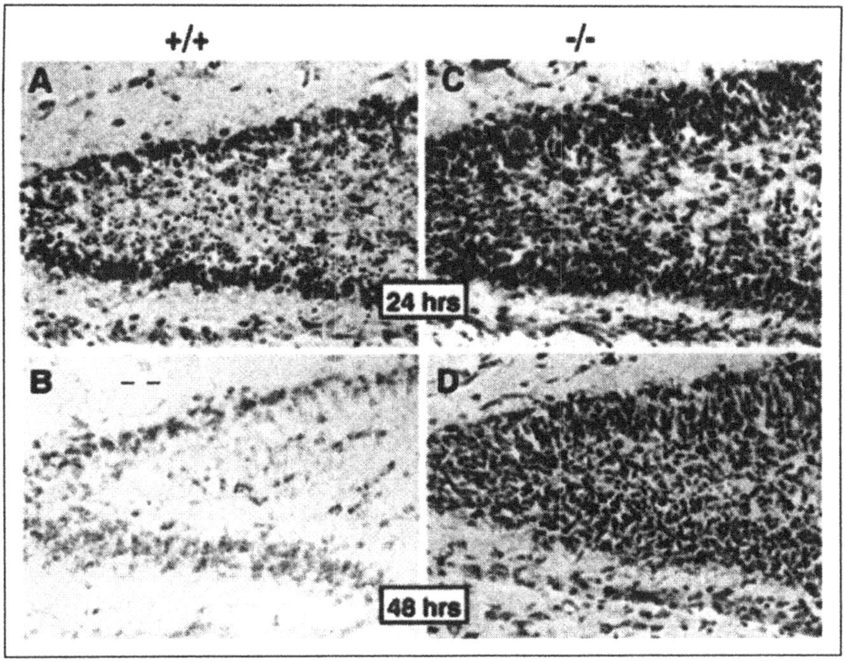

Figure 2. Absence of apoptosis in response to IR in regions of the central nervous system of ATM$^{-/-}$ mice. 24 hrs after ionizing irradiation, the developing hippocampal dentate gyrus of WT mice ($^{+/+}$) show a typical amount of cell death (A) and extensive cell depletion 48 hours after irradiation (B). However, ATM$^{-/-}$ dentate gyrus shows marked resistance to irradiation for up to 48 hours (C and D), with no obvious cell depletion. From Herzog KH, Chong MJ, Kapsetaki M, Morgan JI, McKinnon PJ. Science. 1998; 280(5366):1089-1091. Reprinted with permission from AAAS.

The cloning of the ATM gene in 1995 has triggered an explosion in our understanding of the interconnections between DNA damage checkpoints and repair and apoptosis. It has become clear that the DNA damage signaling network is extremely large and complex. In recent efforts to elucidate the functional mechanisms of ATM and ATR, investigators have identified over 700 proteins.[12] The sheer number and diversity of the substrates thus far identified suggests that the roles of ATM and ATR may be more widespread than previously thought. Many of the substrates are involved in cell cycle arrest while fewer, though still a significant number, are involved with apoptosis. The identification of novel substrates raises interesting connections. Ongoing research promises to clarify the mechanisms of ATM-mediated cell cycle arrest and apoptosis as well as elucidate additional roles in numerous cellular responses.

Acknowledgements

This work was supported by the Ruth L Kirschstein National Research Service Award CA119724 (D Wilsker), the Flight Attendant Medical Research Institute and the National Cancer Institute CA104253 (F Bunz).

References

1. Sherr CJ, Roberts JM. Living with or without cyclins and cyclin-dependent kinases. Genes Dev 2004; 18:2699-2711.
2. Donzelli M, Draetta GF. Regulating mammalian checkpoints through Cdc25 inactivation. EMBO Rep 2003; 4:671-677.

3. Bartek J, Lukas J. Pathways governing G1/S transition and their response to DNA damage. FEBS Lett 2001; 490:117-122.
4. Zhou BB, Elledge SJ. The DNA damage response: Putting checkpoints in perspective. Nature 2000; 408:433-439.
5. Kastan MB, Bartek J. Cell-cycle checkpoints and cancer. Nature 2004; 432:316-323.
6. Kitagawa R, Kastan MB. The ATM-dependent DNA damage signaling pathway. Cold Spring Harb Symp Quant Biol 2005; 70:99-109.
7. Hurley PJ, Bunz F. ATM and ATR: Components of an integrated circuit. Cell Cycle 2007; 6:414-417.
8. Stiff T, Walker SA, Cerosaletti K et al. ATR-dependent phosphorylation and activation of ATM in response to UV treatment or replication fork stalling. EMBO J 2006; 25:5775-5782.
9. Dodson GE, Tibbetts RS. DNA replication stress-induced phosphorylation of cyclic AMP response element-binding protein mediated by ATM. J Biol Chem 2006; 281:1692-1697.
10. Kim ST, Xu B, Kastan MB. Involvement of the cohesin protein, Smc1, in atm-dependent and independent responses to DNA damage. Genes Dev 2002; 16:560-570.
11. Kitagawa R, Bakkenist CJ, McKinnon PJ et al. Phosphorylation of SMC1 is a critical downstream event in the ATM-NBS1-BRCA1 pathway. Genes Dev 2004; 18:1423-1438.
12. Matsuoka S, Ballif BA, Smogorzewska A et al. ATM and ATR substrate analysis reveals extensive protein networks responsive to DNA damage. Science 2007; 316:1160-1166.
13. Matsuoka S, Rotman G, Ogawa A et al. Ataxia telangiectasia-mutated phosphorylates Chk2 in vivo and in vitro. Proc Natl Acad Sci USA. 2000; 97:10389-10394.
14. Gatei M, Sloper K, Sorensen C et al. Ataxia telangiectasia-mutated (ATM) and NBS1-dependent phosphorylation of Chkl on ser-317 in response to ionizing radiation. J Biol Chem 2003; 278:14806-14811.
15. Smits VA. Spreading the signal: Dissociation of Chkl from chromatin. Cell Cycle 2006; 5:1039-1043.
16. Yazdi PT, Wang Y, Zhao S et al. SMC1 is a downstream effector in the ATM/NBS1 branch of the human S-phase checkpoint. Genes Dev 2002; 16:571-582.
17. Xu B, Kim S, Kastan MB. Involvement of Brcal in S-phase and G(2)-phase checkpoints after ionizing irradiation. Mol Cell Biol 2001; 21:3445-3450.
18. Jhanwar-Uniyal M. BRCA1 in cancer, cell cycle and genomic stability. Front Biosci 2003; 8:s1107-1117.
19. Taniguchi T, Garcia-Higuera I, Andreassen PR et al. S-phase-specific interaction of the fanconi anemia protein, FANCD2, with BRCA1 and RAD51. Blood 2002; 100:2414-2420.
20. Taniguchi T, Garcia-Higuera I, Xu B et al. Convergence of the fanconi anemia and ataxia telangiectasia signaling pathways. Cell 2002; 109:459-472.
21. Ahmad SI, Kirk SH, eds. Molecular Mechanisms of Fanconi Anemia. Austin/New York: Landes Bioscience/Springer, 2006.
22. Bartek J, Lukas J. Mammalian G1- and S-phase checkpoints in response to DNA damage. Curr Opin Cell Biol 2001; 13:738-747.
23. Matsuoka S, Huang M, Elledge SJ. Linkage of ATM to cell cycle regulation by the Chk2 protein kinase. Science 1998; 282:1893-1897.
24. Melchionna R, Chen XB, Blasina A et al. Threonine 68 is required for radiation-induced phosphorylation and activation of Cds1. Nat Cell Biol 2000; 2:762-765.
25. Falck J, Mailand N, Syljuasen RG et al. The ATM-Chk2-Cdc25A checkpoint pathway guards against radioresistant DNA synthesis. Nature 2001; 410:842-847.
26. Sorensen CS, Syljuasen RG, Falck J et al. Chkl regulates the S-Phase checkpoint by coupling the physiological turnover and ionizing radiation-induced accelerated proteolysis of Cdc25A. Cancer Cell 2003; 3:247-258.
27. Kastan MB, Zhan Q, el-Deiry WS et al. A mammalian cell cycle checkpoint pathway utilizing p53 and GADD45 is defective in ataxia telangiectasia. Cell 1992; 71:587-597.
28. Canman CE, Lim DS, Cimprich KA et al. Activation of the ATM kinase by ionizing radiation and phosphorylation of p53. Science 1998; 281:1677-1679.
29. Lakin ND, Hann BC, Jackson SP. The ataxia telangiectasia related protein ATR mediates DNA-dependent phosphorylation of p53. Oncogene 1999; 18:3989-3995.
30. Tibbetts RS, Brumbaugh KM, Williams JM et al. A role for ATR in the DNA damage-induced phosphorylation of p53. Genes Dev 1999; 13:152-157.
31. Chehab NH, Malikzay A, Appel M et al. Chk2/hCds1 functions as a DNA damage checkpoint in G(1) by stabilizing p53. Genes Dev 2000; 14:278-288.
32. Shieh SY, Ahn J, Tamai K et al. The human homologs of checkpoint kinases Chkl and Cds1 (Chk2) phosphorylate p53 at multiple DNA damage-inducible sites. Genes Dev 2000; 14:289-300.
33. Canman CE, Wolff AC, Chen CY et al. The p53-dependent G1 cell cycle checkpoint pathway and ataxia telangiectasia. Cancer Res 1994; 54:5054-5058.

34. Xu Y, Yang EM, Brugarolas J et al. Involvement of p53 and p21 in cellular defects and tumorigenesis in Atm$^{-/-}$ mice. Mol Cell Biol 1998; 18:4385-4390.
35. Xie G, Habbersett RC, Jia Y et al. Requirements for p53 and the ATM gene product in the regulation of G1/S and S-Phase checkpoints. Oncogene 1998; 16:721-736.
36. Berkovich E, Monnat RJ Jr, Kastan MB. Roles of ATM and NBS1 in chromatin structure modulation and DNA double-strand break repair. Nat Cell Biol 2007; 9:683-690.
37. You Z, Bailis JM, Johnson SA et al. Rapid activation of ATM on DNA flanking double-strand breaks. Nat Cell Biol 2007; 9:1311-1318.
38. Lim DS, Kim ST, Xu B et al. ATM phosphorylates p95/nbs1 in an S-phase checkpoint pathway. Nature 2000; 404:613-617.
39. Wu X, Ranganathan V, Weisman DS et al. ATM phosphorylation of nijmegen breakage syndrome protein is required in a DNA damage response. Nature 2000; 405:477-482.
40. Zhao S, Weng YC, Yuan SS et al. Functional link between ataxia telangiectasia and nijmegen breakage syndrome gene products. Nature 2000; 405:473-477.
41. Painter RB, Young BR. Radiosensitivity in ataxia telangiectasia: A new explanation. Proc Natl Acad Sci USA 1980; 77:7315-7317.
42. Taylor AM, Harnden DG, Arlett CF et al. Ataxia telangiectasia: A human mutation with abnormal radiation sensitivity. Nature 1975; 258:427-429.
43. Beamish H, Williams R, Chen P et al. Defect in multiple cell cycle checkpoints in ataxia telangiectasia postirradiation. J Biol Chem 1996; 271:20486-20493.
44. Terzoudi GI, Manola KN, Pantelias GE et al. Checkpoint abrogation in G2 compromises repair of chromosomal breaks in ataxia telangiectasia cells. Cancer Res 2005; 65:11292-11296.
45. Bunz F, Dutriaux A, Lengauer C et al. Requirement for p53 and p21 to sustain G2 arrest after DNA damage. Science 1998; 282:1497-1501.
46. Bruno T, De Nicola F, Iezzi S et al. Che-1 phosphorylation by ATM/ATR and Chk2 kinases activates p53 transcription and the G2/M checkpoint. Cancer Cell 2006; 10:473-486.
47. Li S, Ting NS, Zheng L et al. Functional link of BRCA1 and ataxia telangiectasia gene product in DNA damage response. Nature 2000; 406:210-215.
48. Reinhardt HC, Aslanian AS, Lees JA et al. p53-deficient cells rely on ATM- and ATR-mediated checkpoint signaling through the p38MAPK/MK2 pathway for survival after DNA damage. Cancer Cell 2007; 11:175-189.
49. Bree RT, Neary C, Samali A et al. The switch from survival responses to apoptosis after chromosomal breaks. DNA Repair (Amst) 2004; 3:989-995.
50. Chen X, Ko LJ, Jayaraman L et al. p53 levels, functional domains and DNA damage determine the extent of the apoptotic response of tumor cells. Genes Dev 1996; 10:2438-2451.
51. Latonen L, Laiho M. Cellular UV damage responses—functions of tumor suppressor p53. Biochim Biophys Acta 2005; 1755:71-89.
52. Vousden KH. P53: Death star. Cell 2000; 103:691-694.
53. Giaccia AJ, Kastan MB. The complexity of p53 modulation: Emerging patterns from divergent signals. Genes Dev 1998; 12:2973-2983.
54. Herzog KH, Chong MJ, Kapsetaki M et al. Requirement for atm in ionizing radiation-induced cell death in the developing central nervous system. Science 1998; 280:1089-1091.
55. Baker SJ, McKinnon PJ. Tumour-suppressor function in the nervous system. Nat Rev Cancer 2004; 4:184-196.
56. Stagni V, di Bari MG, Cursi S et al. ATM kinase activity modulates fas sensitivity through the regulation of FLIP in lymphoid cells. Blood 2008; 111:829-837.
57. Lee Y, McKinnon PJ. Responding to DNA double strand breaks in the nervous system. Neuroscience 2007; 145:1365-1374.
58. Soares HD, Morgan JI, McKinnon PJ. Atm expression patterns suggest a contribution from the peripheral nervous system to the phenotype of ataxia telangiectasia. Neuroscience 1998; 86:1045-1054.
59. McKinnon PJ. ATM and ataxia telangiectasia. EMBO Rep 2004; 5:772-776.

Chapter 8

Ataxia Telangiectasia: An Oxidative Stress-Related Disease

Giovanni Pagano,* Paolo Degan and Giuseppe Castello

Abstract

Multi-factorial information is available on Ataxia Telangiectasia (AT) phenotype associated with oxidative stress. First, the protein encoded by the AT gene (ATM) is involved in both the regulation of responses to double strand DNA breaks and of response to oxidative stress. Moreover, an established collection of literature points to the roles of oxidative stress in AT that can be summarized as follows: (i) increased sensitivity of AT cells to DNA damaging agents via free radical mechanisms; (ii) increased sensitivity of AT cells to inflammatory cells and to ROS; (iii) protective effects of synthetic antioxidants in Atm knock-out mice; (iv) regulation of oxidative stress response in cell proliferation vs apoptosis, (v) abnormal levels of oxidative stress parameters in body fluids and blood cells from AT patients and (vi) impairment of mitochondrial functions and structure. Oxidative stress may thus be envisaged as a major phenomenon in AT's clinical phenotype.

Introduction

Ataxia telangiectasia is an autosomal recessive syndrome characterized by cerebellar ataxia and neuromotor impairment, conjunctival and cutaneous telangiectasias, growth retardation, immunodeficiency, sterility, genomic instability, progeric skin and hair changes, propensity to lymphoreticular malignancies and sinopulmonary infections.[1-5] Chromosomal instability is associated with a high sensitivity to ionizing radiation and radiomimetic chemicals.[6] The AT gene-encoded protein, ATM is a protein phosphatidylinositol-3' (PI-3) kinase,[1] involved in the multifunctional regulation of cellular responses to double strand DNA breaks and in other processes that maintain cellular homeostasis, also suggesting implications of oxidative stress in the AT phenotype.[7-8]

Starting from pioneering reports by Shiloh et al[6,9] the relationships between AT phenotype and oxidative stress have been assessed in a number of studies with a steadily growing trend, as shown in (Fig. 1). These relationships rely on in vivo and in vitro human cell data,[10-19] animal studies[20-27] and molecular evidence[19,28-34] as summarized in Table 1.

A Composite Clinical Phenotype: Seeking a Unified Frame

The redox-related information on AT phenotype includes multi-factorial implications of oxidative stress in AT's clinical features and in disease progression. A multi-factorial collection of evidence points to an endogenous pro-oxidant state in AT phenotype, suggesting that the ATM protein kinase is involved in regulating cellular redox homeostasis and in modulating the expression of proteins with antioxidant functions.[7-8,30-33] The most commonly observed clinical features or AT-related complications that have been well-recognized for their relationships with oxidative stress include: (a) lymphoproliferative disorders;[2,5,35] (b) neurological impairments;[4,36]

*Corresponding Author: Giovanni Pagano—Italian National Cancer Institute, CROM; I-83013 Mercogliano (AV), Italy. Email: gbpagano@tin.it

Molecular Mechanisms of Ataxia Telangiectasia, edited by Shamim I. Ahmad.

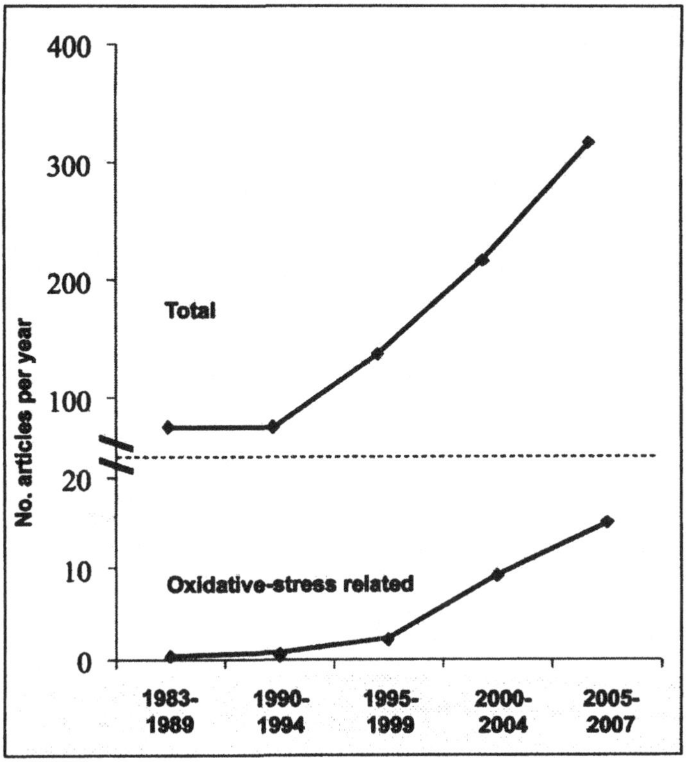

Figure 1. Scores of publications referred to oxidative stress or DNA repair referred to the main features of AT phenotype starting from 1983, the year of Shiloh et al paper.[9] Sources: MedLine, February 2008 and authors' archives.

Table 1. Outline of the evidence for the involvement of oxidative stress in ataxia telangiectasia

Endpoints	Materials	References
↑ Sensitivity to DNA damaging agents via free radical mechanisms	Human AT cells	6,9
↑ Sensitivity to inflammatory cells and to ROS	Human AT cells	15,16
↑ SOD and catalase activity	AT patient erythrocytes	10
Regulation of oxidative stress response in cell proliferation vs apoptosis	Human AT cells	15,26-30
↓↑ Oxidative stress parameters	Body fluids and blood cells from AT patients	11-13
Low molecular weight antioxidants as protection factors	$Atm^{-/-}$ mice	18-23
Alterations in redox state	Brains from $Atm^{-/-}$ mice	24,25

(c) immunodeficiency[4,37] (d) progeroid phenotype,[4,38] and Type II diabetes mellitus.[39-41] Thus, a key role may be attributed to oxidative stress in the AT clinical and cellular phenotypes, including the progression to malignancies, as well as in some of the clinical complications and features associated with AT (Fig. 2).

AT and Mitochondrial Dysfunction

A relevant role for mitochondrial dysfunction in AT phenotype has been suggested in a recent study by Ambrose et al[42] providing evidence for abnormalities in the structural organization of mitochondria and for a decreased membrane potential in AT cells' mitochondria versus mitochondia from control cells. In addition, the levels of basal expression of several nuclear DNA-encoded oxidative damage responsive genes whose proteins are targeted to the mitochondria—polymerase gamma, mitochondrial topoisomerase I, peroxiredoxin 3 and manganese superoxide dismutase (Mn-SOD), all these were found to be elevated in AT cells. Consistent with these results, these authors also found that, overall mitochondrial respiratory activity was diminished in AT cells compared to wild type cells.[42] Treating AT cells with the antioxidant, alpha lipoic acid (ALA), restored mitochondrial respiration rates to levels approaching those of wild type. When wild type cells were transfected with ATM-targeted siRNA, a small but significant reduction was found in the respiration rates of mitochondria. Moreover, mitochondria in AT cells, induced to stably express full-length ATM, exhibited respiration rates approaching those of wild type cells. Together, this study provides evidence for an intrinsic mitochondrial dysfunction in AT cells, implicating a requirement for ATM in the regulation of mitochondrial function.[42]

Another study has shown increased accumulation of the common mitochondrial DNA deletion in AT cells versus mitochondrial DNA in controled cells, as both SV40 transformed and nontransformed AT cell lines showed significant induction of the common mitochondrial deletion.[43]

Mitochondrial dysfunction has been reported as an effector in the aging process and has been shown to be an important contributing factor to the disease phenotypes associated with many progressive neurodegenerative disorders, such as Alzheimer's disease, Parkinson's disease, Huntington's disease, amyotrophic lateral sclerosis (ALS) and Friedreich ataxia.[44-45] Together,

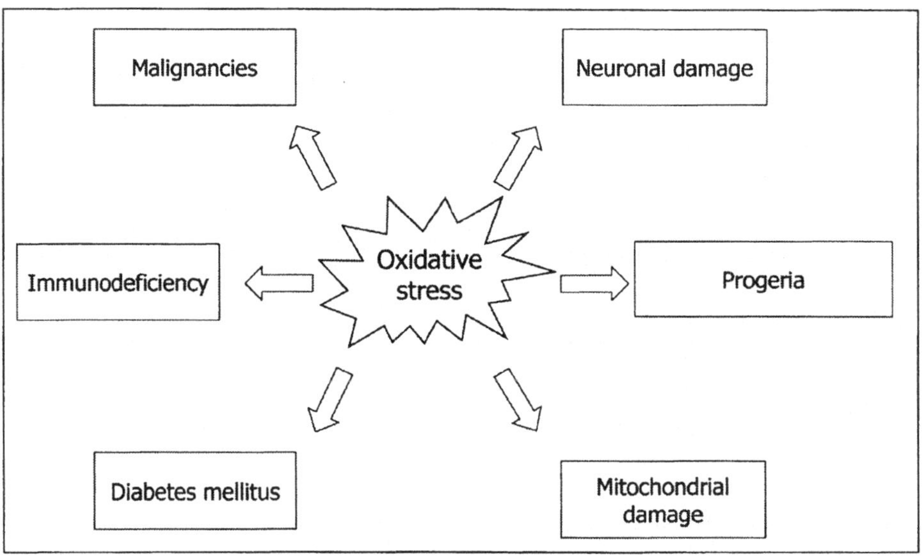

Figure 2. Oxidative stress as a central event related to the major clinical and cellular abnormalities observed in AT.

mitochondrial dysfunction may be regarded as playing a key-role contributing to the continuous oxidative stress of AT cells as well.

The majority of ROS formation in eukaryotic cells occurs in mitochondria as by-products during the generation of adenosine triphosphate (ATP), through the processes of electron transport chain/ oxidative phosphorylation system (ETC/OXPHOS) and of the tricarboxylic acid cycle (TAC).[46] These processes display a limited efficiency, as it is estimated that 1-5% of electrons flowing through the ETC system are donated to molecular oxygen forming superoxide ($O_{2-}^{\bullet}$).[46-49] In turn, $^{\bullet}O_{2-}$ can be removed by manganese superoxide dismutase (MnSOD) to form hydrogen peroxide (H_2O_2) which can readily diffuse into the cytosol and nucleus and, in the presence of transition metals, resulting in the production of highly reactive hydroxyl radicals ($OH^{\bullet}$).[49]

A recent report by Berni et al has shown a protective effect of l-carnitine on tert-butyl-hydroperoxide-induced DNA damage and on chromosomal instability.[18] This finding may be related to the well-established proctective role of l-carnitine on mitochondrial functions.[49]

Prospects of Chemoprevention Trials

The growing evidence for the involvement of oxidative stress in AT, like in some other genetic diseases, has raised the prospect of counteracting in vivo prooxidant states by targeted antioxidant administration aimed at improving disease course by decreasing oxidative stress-associated complications.[50]

For a rare disease as AT, insuperable obstacles may currently be faced with due to the limited numbers of patients susceptible to be recruited, jeopardizing the best efforts in drawing statistically sound conclusions from too small scale clinical studies. As a possible means for overcoming the restraint imposed by the scarcity of human patients, studies have been designed by utilizing knock-out mice for AT.[20-25]

In either animal studies or in the prospect of future chemoprevention trials on human AT patients, endpoints should not be confined to changes in lifespan or in some pathological outcomes such as cancer incidence, or disease-specific complications. Rather, a defined set of oxidative stress parameters should be measured, by verifying any baseline alterations vs control values, as a prerequisite for undertaking any chemopreventive treatment. Thereafter, any changes in the in vivo prooxidant states should be monitored at defined time intervals in blood, urine and in the relevant target tissues, e.g., brain using AT knock-out mice. Thus the efficacy, if any, of the tested antioxidants, in counteracting the initial prooxidant state, should allow us to combine the data of a biochemical compensation vs a clinical improvement.[50]

Another major issue in study design should rely on the choice of the most appropriate antioxidants. These should be selected in view of the baseline redox imbalances and of the biochemical targets to be achieved. The choice of antioxidants to be used in either animal studies or, even more so, in clinical trials should preferentially be confined to molecules being already approved for safe human use with minimal showing no adverse side effects as, e.g., N-acetylcysteine.[23-24] Despite the successful test results using tempol in $Atm^{-/-}$ mice,[20,22,25] this chemical cannot, to the best our knowledge, be applied on humans due to the current lack of safety information.

Conclusion

Available convincing evidence leads to the assessment that the multiple outcomes in AT phenotype respond to the scenarios of an inborn pro-oxidant state that is likely to be originated by mitochondrial dysfunction. We may envisage that the AT-associated clinical pattern, as a set of endogenous events may result in a combined series of pathologies and/or of abnormalities in redox endpoints.

In conclusion, the current state-of-the-art may prompt the medical community to recognize that AT may be regarded as an oxidative stress-related disorder. This definition might lead to a unifying interpretation of AT, with possible implications on clinical study design towards an advancement in patients' management.

References

1. Chun HH, Gatti RA. Ataxia telangiectasia, an evolving phenotype. DNA Repair (Amst) 2004; 3:1187-1196.
2. Perlman S, Becker-Catania S, Gatti RA. Ataxia telangiectasia: diagnosis treatment. Semin Pediatr Neurol 2003; 10:173-182.
3. Rosen FG, Eibl M, Roifman C et al. Primary Immunodeficiency Diseases. Report of an IUIS scientific committee. International Union of Immunological Societies. Clin Exp Immunol 1999; 118(Suppl 1): 1-28.
4. Subba Rao K. Mechanisms of disease: DNA repair defects and neurological disease. Nat Clin Pract Neurol 2007; 3:162-72.
5. Gumy-Pause F, Wacker P, Sappino AP. ATM gene and lymphoid malignancies. Leukemia 2004; 18:238-242.
6. Shiloh Y, Parshad R, Frydman M et al. G2 chromosomal radiosensitivity in families with ataxia telangiectasia. Hum Genet 1985; 84:15-18.
7. Barzilai A, Rotman G, Shiloh Y. ATM deficiency and oxidative stress: a new dimension of defective response to DNA damage. DNA Repair (Amst) 2002; 1:3-25.
8. Watters DJ. Oxidative stress in ataxia telangiectasia. Redox Rep 2003; 8:23-29.
9. Shiloh Y, Tabor E, Becker Y. Abnormal response to ataxia telangiectasia cells to agents that break the deoxyribose moiety of DNA via a targeted free radical mechanism. Carcinogenesis 1983; 4:1317-1322.
10. Aksoy Y, Sanal O, Metin A et al. Antioxidant enzymes in red blood cells and lymphocytes of ataxia telangiectasia patients. Turk J Pediatr 2004; 46:204-207.
11. Reichenbach J, Schubert R, Schwan C et al. Anti-oxidative capacity in patients with ataxia telangiectasia. Clin Exp Immunol 1999; 117: 535-539.
12. Reichenbach J, Schubert R, Schindler D et al. Elevated oxidative stress in patients with ataxia telangiectasia. Antioxid Redox Signal 2002; 4:465-469.
13. Degan P, d'Ischia M, Pallardó FV et al. Glutathione levels in blood from ataxia telangiectasia patients suggest in vivo adaptive mechanisms to oxidative stress. Clin Biochem 2007; 40:666-670.
14. Barlow C, Dennery PA, Shigenaga MK et al (1999) Loss of the ataxia telangiectasia gene product causes oxidative damage in target organs. Proc Natl Acad Sci USA 1999; 96:9915–9919.
15. Shackelford RE, Innes CL, Sieber SO et al. The Ataxia telangiectasia gene product is required for oxidative stress-induced G1 and G2 checkpoint function in human fibroblasts. J Biol Chem 2001; 276:21951-21959.
16. Ward AJ, Olive PL, Burr AH et al. Response of fibroblast cultures from ataxia telangiectasia patients to reactive oxygen species generated during inflammatory reactions. Environ Mol Mutagen 1994; 24:103-111.
17. Yi M, Rosin MP, Anderson CK. Response of fibroblast cultures from ataxia telangiectasia patients to oxidative stress. Cancer Lett 1990; 54:43-50.
18. Berni A, Meschini R, Filippi S et al. l-Carnitine enhances resistance to oxidative stress by reducing DNA damage in Ataxia telangiectasia cells. Mutat Res 2008; 650:165-174.
19. Valerie K, Yacoub A, Hagan MP et al. Radiation-induced cell signaling: inside-out and outside-in. Mol Cancer Ther 2007; 6:789-801.
20. Browne SE, Roberts LJ 2nd, Dennery PA et al. Treatment with a catalytic antioxidant corrects the neurobehavioral defect in ataxia telangiectasia mice. Free Radic Biol Med 2004; 36:938-942.
21. Erker L, Schubert R, Elchuri S et al. Effect of the reduction of superoxide dismutase 1 and 2 or treatment with alpha tocopherol on tumorigenesis in Atm-deficient mice. Free Radic Biol Med 2006; 41:590-600.
22. Gueven N, Luff J, Peng C et al. Dramatic extension of tumor latency and correction of neurobehavioral phenotype in Atm-mutant mice with a nitroxide antioxidant. Free Radic Biol Med 2006; 41:992-1000.
23. Reliene R, Fischer E, Schiestl RH. Effect of N-acetyl cysteine on oxidative DNA damage and the frequency of DNA deletions in atm-deficient mice. Cancer Res 2004; 64:5148-5153.
24. Reliene R, Fleming SM, Chesselet MF et al. Effects of antioxidants on cancer prevention and neuromotor performance in Atm deficient mice. Food Chem Toxicol 2008; 46:1371-1377.
25. Schubert R, Erker L, Barlow C et al. Cancer chemoprevention by the antioxidant tempol in Atm-deficient mice. Hum Mol Genet 2004; 13:1793-1802.
26. Chen P, Peng C, Luff J et al. Oxidative stress is responsible for deficient survival and dendritogenesis in Purkinje neurons from ataxia telangiectasia mutated mutant mice. J Neurosci 2003; 23:11453-11460.
27. Kamsler A, Daily D, Hochman A et al. Increased oxidative stress in ataxia telangiectasia evidenced by alterations in redox state of brains from Atm-deficient mice. Cancer Res 2001; 61:1849-1854.
28. Bagley J, Singh G, Iacomini J. Regulation of oxidative stress responses by ataxia telangiectasia mutated is required for T-cell proliferation. J Immunol 2007; 178:4757-4763.

29. Kobayashi M, Ono H, Mihara K et al. ATM activation by a sulfhydryl-reactive inflammatory cyclopentenone prostaglandin. Genes Cells 2006; 11:779-789.
30. Tanaka T, Halicka HD, Huang X et al. Constitutive histone H2AX phosphorylation and ATM activation, the reporters of DNA damage by endogenous oxidants. Cell Cycle 2006; 5:1940-1945.
31. Dong Z, Tomkinson AE. ATM mediates oxidative stress-induced dephosphorylation of DNA ligase IIIalpha. Nucleic Acids Res 2006; 34:5721-5729.
32. Yan M, Shen J, Person MD et al. Endoplasmic reticulum stress and unfolded protein response in Atm-deficient thymocytes and thymic lymphoma cells are attributable to oxidative stress. Neoplasia 2008; 10:160-167.
33. Taylor A, Shang F, Nowell T et al. Ubiquitination capabilities in response to neocarzinostatin and H_2O_2 stress in cell lines from patients with ataxia telangiectasia. Oncogene 2002; 21:4363-4373.
34. Das KC, Dashnamoorthy R. Hyperoxia activates the ATR-Chk1 pathway and phosphorylates p53 at multiple sites. Am J Physiol Lung Cell Mol Physiol 2004; 286:L87-97.
35. Tohyama Y, Takano T, Yamamura H. B-cell responses to oxidative stress. Curr Pharm Des 2004; 10:835-839.
36. Reynolds A, Laurie C, Mosley RL et al. Oxidative stress and the pathogenesis of neurodegenerative disorders. Int Rev Neurobiol 2007; 82:297-325.
37. Steiner J, Haughey N, Li W et al. Oxidative stress and therapeutic approaches in HIV dementia. Antioxid Redox Signal 2006; 8:2089-2100.
38. Pagano G, Zatterale A, Degan P et al. In vivo prooxidant state in Werner syndrome patients. Free Radic Res 2005; 39:529-533.
39. Pagano G, Degan P, d'Ischia M et al. Oxidative stress as a multiple effector in Fanconi anaemia clinical phenotype. Eur J Haematol 2005; 75:93-100.
40. Lee HB, Seo JY, Yu MR et al. Radical approach to diabetic nephropathy. Kidney Int Suppl 2007; 106:S67-70.
41. Ristow M. Neurodegenerative disorders associated with diabetes mellitus. J Mol Med 2004; 82:510-529.
42. Ambrose M, Goldstine JV, Gatti RA. Intrinsic mitochondrial dysfunction in ATM-deficient lymphoblastoid cells. Hum Mol Genet 2007; 16:2154-2164.
43. Prithivirajsingh S, Story MD, Bergh SA et al. Accumulation of the common mitochondrial DNA deletion induced by ionizing radiation. FEBS Lett 2004; 571:227-232.
44. Kidd PM. Neurodegeneration from mitochondrial insufficiency: nutrients, stem cells, growth factors and prospects for brain rebuilding using integrative management. Altern Med Rev 2005; 10:268-293.
45. Kwong JQ, Beal MF, Manfredi G. The role of mitochondria in inherited neurodegenerative diseases. J Neurochem 2006; 97:1659-1675.
46. Schapira AH. Mitochondrial disease. Lancet 2006; 368:70-82.
47. Schon EA, Manfredi G. Neuronal degeneration and mitochondrial dysfunction. J Clin Invest 2003; 111:303-312.
48. Zeevalk GD, Bernard LP, Song C et al. Mitochondrial inhibition and oxidative stress: reciprocating players in neurodegeneration. Antioxid Redox Signal 2005; 7:1117-1139.
49. Steiber A, Kerner J, Hoppel C. Carnitine: a nutritional, biosynthetic and functional perspective. Mol Aspects Med 2004; 25:455-473.
50. Lloret A, Calzone R, Dunster C et al. Different patterns of in vivo prooxidant states in a set of cancer- or ageing-related genetic diseases. Free Radic Biol Med 2007; 44:495-503.

Chapter 9

Oncogenesis in Ataxia Telangectasia:
Roles of ATM, p53, NF-κB and DDE Recombination Pathogenesis

David H. Dreyfus*

Abstract

The mechanistic basis of ATM (Ataxia Telangectasis Mutated) protein interactions with DNA hairpin and related double stranded DNA breaks, generated through V(D)J recombination, has been the subject of numerous recent experimental reports and reviews. The novel focus of this review is the potential sources of unresolved or aberrant hairpin formation resulting in oncogenic DNA recombination events rather than the mechanistic role of ATM in the process of V(D)J recombination. Experimental evidence is reviewed regarding the critical role of ATM in sensing and repair of pathogenic hairpin V(D)J-like recombination events occurring V(D)J RSS (Recombination Signal Sequences) as well as at endogenous transposable element termini and other cryptic V(D)J RSS like sequences in vertebrates. Evidence from recombination associated with transposons in the invertebrate nematode *Caenorhabditis elegans*, an organism that lacks ATM is also reviewed as a model system for understanding oncogenic consequences of loss of ATM in vertebrates. The tumor suppressor p53 is required for the function of ATM. Since it is apparent that p53 is phylogenetically related to and functionally regulated by coregulatory pathway of NF-κB transcription factors, potentially the effector functions of ATM could be blocked not only through loss of the ATM gene but also by alterations in p53 and NF-κB expression. This model suggests the possibility that genomic stress, due to the activation of endogenous transposons and episomal viral pathogens such as Epstein-Barr Virus (EBV), could specifically trigger complex interacting pathogenic pathways involving NF-κB and p53 transcription factors interfering with ATM and a phenotype analogous to loss of ATM in cells with functional ATM protein.

ATM, V(D)J Recombination and V(D)J Recombination Pathogenesis

Molecular dissection of ATM protein,[1] over the past decade, has clarified both the defects in immune repertoire development and DNA repair resulting from loss of ATM protein.[2-11] Both defects in immune repertoire and DNA repair are related to recognition and repair of a specific type of DNA break termed a "hairpin"[12] generated by the process of V(D)J recombination.[13-16]

V(D)J recombination of immunoglobulin and T-cell receptor genes initiated by the RAG proteins RAG-1 and RAG-2 and subsequent processing by additional proteins, DNApk, Ku and Artemis is required for generation of the acquired B and T lymphocyte repertoire as reviewed elsewhere.[17-19] Accumulation of incorrect recombination joints during V(D)J recombination has been termed "V(D)J recombination pathogenesis" postulated to play a role in oncogenesis through

*David H. Dreyfus—Yale SOM and Founder, Keren Pharmaceutical, 488 Norton Parkway, New Haven, CT 06511, USA. Email: dhdreyfus@pol.net

Molecular Mechanisms of Ataxia Telangiectasia, edited by Shamim I. Ahmad.

DNA translocations and other illegitimate recombination events that bring oncogenes into altered regulatory sites or cause abnormal oncogenic fusion proteins.[17,20-24]

ATM in turn senses the presence of unresolved hairpins and arrests cell growth through the tumor suppressor p53 until either the hairpin is repaired or the cell undergoes death through apoptosis.[11,25,26] Since the primary mechanism of V(D)J recombination is breakage and joining of DNA or other nucleic acids, it is essential that this process is highly controlled to prevent potentially oncogenic incorrect joining events.

Phenotype of ATM Deficiency Confirms a Primary Role of the Protein in Detecting and Signaling Presence of Potentially Oncogenic Hairpin Structures

The most plausible explanation for the observation that ATM deficiency causes defects in V(D)J repertoire and also in repair of DNA breakage from other types of DNA damage such as class switch recombination and radiation damage is that the ATM protein recognizes both hairpin DNA structures due to V(D)J recombination and some other types of DNA damage due to oxidative stress or radiation.[10,16] Interestingly, recent experimental observations suggests that ATM folds nonhairpin like DNA breakage termini into hairpin like intermediates in order to recognize these nonhairpin structures through the hairpin resolution pathway.[15] Experimental evidence also confirms that loss of ATM results in a partial defect in immune repertoire as partial V(D)J recombination products, unprocessed hairpins accumulate and are only slowly processed into mature immunoglobulin and T-cell receptor joints.[4,5,9,27]

ATM has a direct role in binding hairpin structures specific to the V(D)J recombination process and other nonhairpin DNA recognition and repair through interaction with Artemis, but is not required for generation of the hairpins or initial processing events mediated by the RAG proteins, Artemis and DNApk. Loss of the ATM protein is not identical to the phenotype resulting from either loss of the RAG proteins, Artemis or DNApk.[28]

Based upon these observations, increased oncogenic recombination events during V(D)J recombination, for example from V(D)J RSS or cryptic V(D)J RSS-like sequences within and flanking oncogenes could directly cause the increased risk of malignancy in AT. Alternatively, the immunodeficiency of the ATM phenotype could indirectly lead to decreased surveillance of malignant cells due to decreased V(D)J recombination generated repertoire.

V(D)J Like Recombination Signals at Sites of V(D)J Recombination Pathogenesis

Central to V(D)J recombination enzymology is a magnesium ion binding catalytic site composed of acidic Aspartate (D) and Glutamate (E) amino acids formed by the RAG (Recombination Activating Gene) encoded proteins termed RAG-1 and RAG-2. The DDE site is shared with mobile DNA sequences termed transposons, retroviral integrases as well as other nucleic acid processing proteins related to RNAse H.[29] The DDE site in the RAG proteins is the active site of DNA breakage, hairpin formation and rejoining and can cause transposition of sequences resembling V(D)J RSS in vitro,[17,22,30-34,35] as well as generation of the immune repertoire in vivo.

A variety of experimental systems have been devised to look for aberrant recombination events related to V(D)J like sequences in vivo, both in AT due to genetic defects in ATM and also less frequently in cells from patients with apparently normal ATM function.[6,36-38] Even in the absence of the RAG proteins, V(D)J like recombination errors are detected[39] so the process of V(D)J recombination pathogenesis must involve other DDE like recombinases, as yet not discovered.

The remainder of this review will describe and discuss the role of cryptic V(D)J-like sequences in human genome and recombinases encoded by endogenous and episomal elements, combining aspects of these two pathways into a model of V(D)J recombination pathogenesis in cells with apparently normal ATM function. In this model, defects due to genetic loss of ATM in AT can provide insights into oncogenesis in cells with normal ATM function during periods of genomic stress, or oncogenic viral replication.

Observations of Recombination Events Associated in the Invertebrate *C. elegans* as a Model of V(D)J Recombination Pathogenesis

The importance of gatekeeper genes such as ATM and ATM-activated p53 is magnified when one realizes that the human genome is littered with cryptic V(D)J like sequences in common genes and copies of transposons potentially capable of providing sites for incorrect V(D)J recombination and hairpin formation.[40] As shown in Figure 1A, the human immunoglobulin and T-cell termini involved in V(D)J recombination are related not only to invertebrate Tc elements but only to widely dispersed endogenous human elements related to Tc transposons. Representative members of these endogenous elements termed mariner, tigger and merlin are shown in Figure 1A.

As shown in Figure 1B, these Tc elements reproduce and move in the genome through a "cut and paste" mechanism.[40] Many of these transposons show evidence of activity in transposition and DNA breakage events in the vertebrate genome in the fairly recent past and the activity of these sequences in somatic tissues currently is not well characterized. In addition to V(DJ RSS, endogenous transposon termini may provide sites of DNA breakage and hairpin formation contributing to oncogenesis both in patients with AT and in patients without AT during periods of genomic stress or viral infection.

It has been previously proposed by the author that analysis of recombination events due to Tc elements in invertebrate organisms such as *C. elegans* that lack V(D)J recombination and vertebrate regulatory controls of the V(D)J recombination pathway such as ATM may provide insights into V(D)J recombination pathogenesis in vertebrates, for example in Ataxia telangiectasia (AT).[40] In invertebrate species such as *C. elegans* that lack ATM, analysis of recombination within and flanking Tc elements whose termini resemble V(D)J RSS illustrates that transposon like termini are capable of generating complex recombination products through what appears to be a process of internal excision events (Fig. 2).

Although it is often assumed that Tc like transposons are extinct in the germ line of higher vertebrates such as humans, this assumption is incorrect.[40-46] Activity of Tc-like elements in the germ-line appears to have played a role in differentiation of primate species in the fairly recent past. Tc like transposons also contribute to protein evolution in vertebrates through generation of fusion proteins between transposon encoded proteins and proteins at insertion sites. Another unknown area is the activity of Tc like transposons in vertebrate somatic tissues.

Transposons Gone Wild: A Putative Role for Genomic Stress Bypassing ATM through Interactions between Tumor Suppressor p53 and NF-κB Transcription Factors

Experimental systems confirm that the ATM protein is a critical link in the "taming" of the V(D)J recombinase from its original "wild" origins as a transposase and suggest that loss of ATM may also be useful for providing additional insights into V(D)J recombination pathogenesis and cancer. In addition to the well characterized activity of the "tamed" RAG transpose, it seems very likely that a large number of "untamed" or "wild" transposons and transposases are present in the human genome. Left unregulated by ATM as in the *C. elegans* model system, certainly the enormous number of Tc like elements in the human genome would cause genomic fragmentation and chaos typical of the cancer cell "mutator" phenotype.

As summarized in the preceding sections, ATM serves to link recognition and processing of DNA cleaving and processing enzymes in V(D)J recombination such as the RAG proteins and the hairpin opening protein denoted Artemis with effector molecules such as the tumor suppressor denoted p53. Whatever the mechanism of hairpin generation and detection by ATM, following detection of these hairpin or hairpin-like DNA breakage intermediates, ATM activates other gatekeeper proteins causing cell cycle arrest until the defect is repaired. If the hairpin or other breakage intermediate is not repaired, then the defective cell is eliminated by p53 tumor suppressor activating cellular apoptosis.[11,3,47,48]

The author has previously noted that, in vertebrate cells, genomic stress such as transposon activation and/or oncogenic viral infections may contribute to oncogenesis by bypassing DNA repair pathways regulated by ATM, particularly the p53 tumor suppressor.[40,49] As shown in Figure

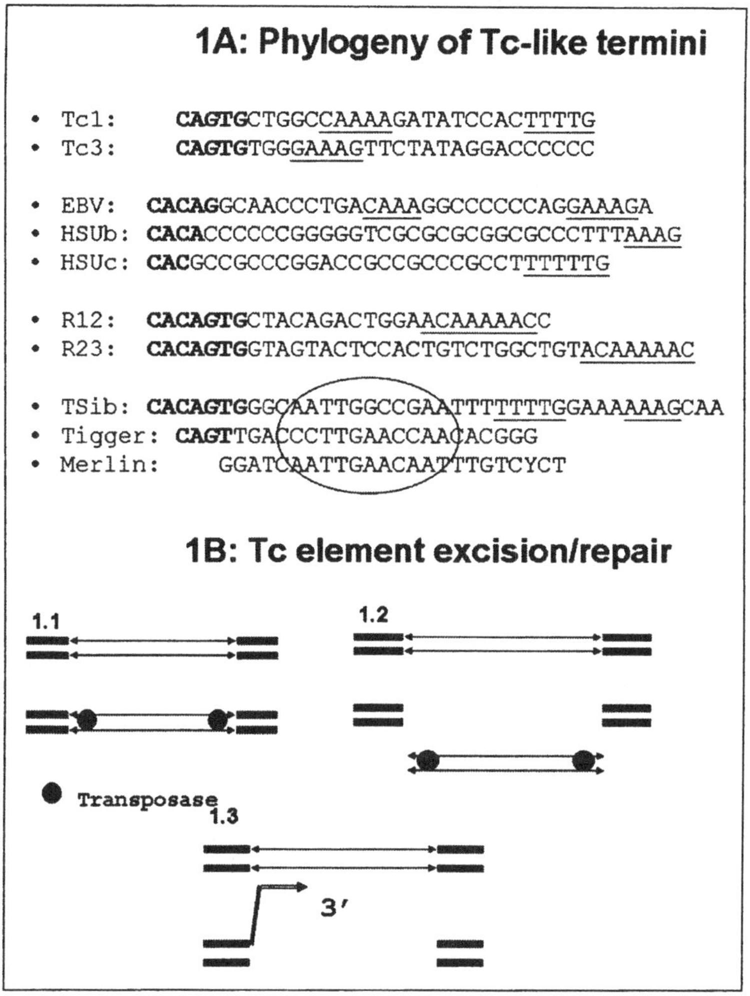

Figure 1. A) Termini of diverse DNA sites recognized by DDE enzymes including human endogenous elements transib, merlin and tigger shows that the endogenous human elements present in large numbers in the human genome share regions of conserved similarity with other mobile DNA sequences including invertebrate Tc elements (Tc1, Tc3), V(D)J RSS heptamers (R12, R23 denotes V(D)J RSS 12 and 23 bp spacers) and the termini of herpes viruses (EBV and HSV). These human endogenous elements share only faint similarities at the immediate cleavage sites (bold face type) and internal regions (circled) suggesting that many other cryptic V(D)J RSS like signals in the genome could be targets of DDE enzymes encoded by these transposons in addition to the RAG proteins. Some endogenous mobile elements and herpes termini appear to have A/T rich regions corresponding the V(D)J RSS nonamer (underlined) while other elements lack these regions proposed to interact with DNA bending proteins B) Mechanism of Tc element excision and transposition by DDE recombinases involves excision of the elements and reinsertion into the genome with repair of the empty site either by replicative copy of the filled site from the sister chromatid generating a filled donor site (shown), or by host enzyme repair pathways not involving the sister chromatid leading to an empty donor site (not shown).[40]

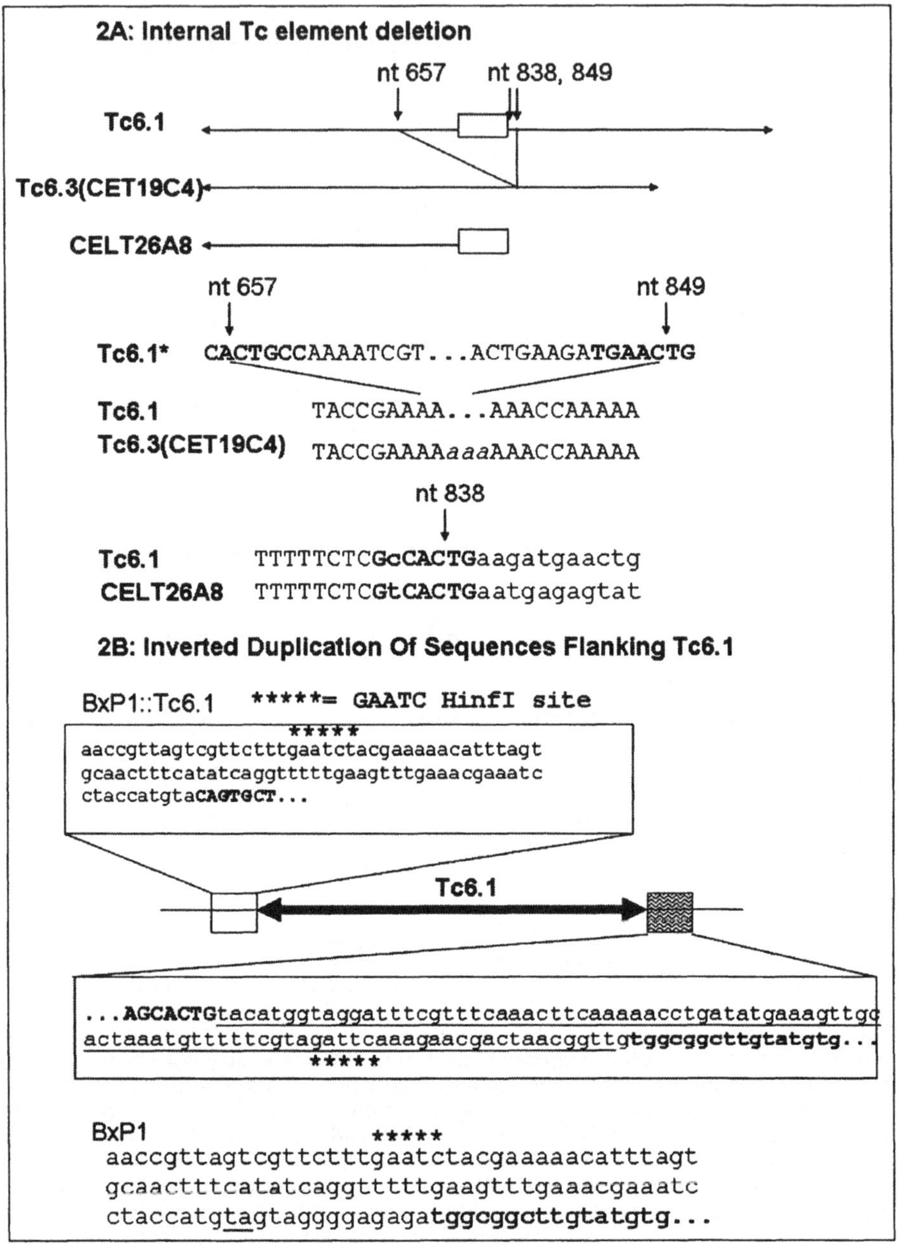

Figure 2. A) Internal deletions of Tc elements in *C. elegans* occur at sequences resembling but not identical to the element termini as described previously.[40] B) Target site inversion and duplication flanking Tc element in *C. elegans* typical of large scale inversions and duplications mediated by DDE enzymes such as transposases as described previously.[40] Complex inversion and duplication of sequences flanking DNA transposons may involve intermediate structures resembling DNA hairpins as discussed in more detail described elsewhere.[40]

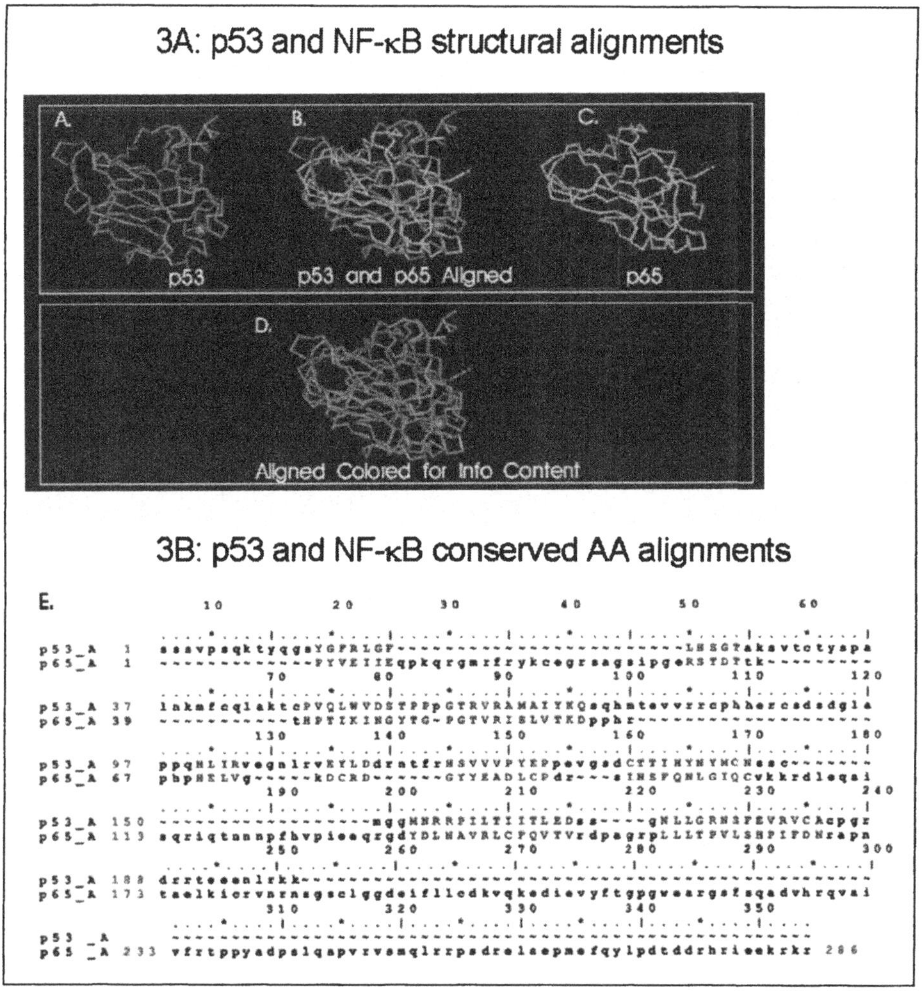

Figure 3. Evidence that p53 and NF-κB share common phylogeny.[49]

3A,B, the p53 protein is in fact a distant relative of the family of NF-κB transcription factors. Remarkably, regulatory factors binding to p53 can bind to NF-κB transcription factors, and conversely regulatory factors binding to NF-κB transcription factors can bind to p53 (Fig. 4). As shown in Figure 5, based on these observations, the author has proposed that p53 and NF-κB transcription factors are descendants of a primordial transcription factor termed "Proto-p53/ NF-κB" regulating both cellular death and proliferation.[49]

It is plausible that there is a threshold effect due to the cross-regulation of p53 and NF-κB transcription factors responding to cell or genomic stress induced by DNA breakage.47,50,51 While some NF-κB transcription may be required for both ATM and p53 expression and function, as NF-κB transcription factors increase, they may be expected to progressively block p53 functions through complex coregulatory pathways. For example, the author has demonstrated that with increasing expression of the NF-κB transcription inhibitor IκBα, p53 function is inactivated and that this effect of IκBα upon p53 localizes to ankyrin-like regions of IκBα that may be shared with viral regulatory proteins (Fig. 5).[49] IκBα levels in turn are increased through a feedback loop by

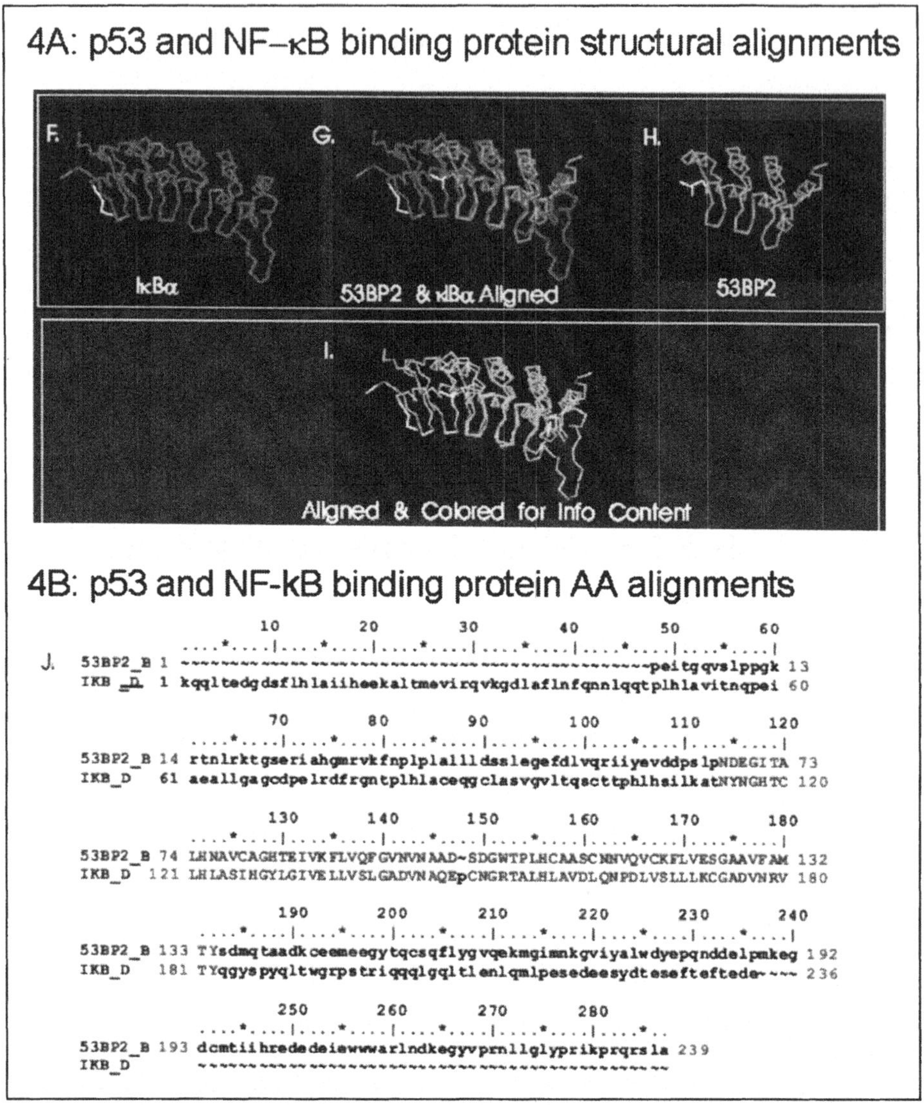

Figure 4. Showing that factors regulating p53 and NF-κB share common phylogeny as described previously.[49]

NF-κB transcription. Thus at high levels of genomic stress and NF-κB transcription, ATM function would be blocked indirectly through loss of p53 function, causing an AT like phenotype even in the presence of functional ATM protein. The result would be oncogenic events resembling the V(D)J recombination pathogenesis evident in AT.

Interactions between p53, NF-κB and ATM could be particularly devastating in the case of oncogenic viral infections such as infection with Epstein-Barr Virus (EBV), an oncogenic virus that may encode a protein(s) that modulate both NF-κB transcription factors[49] and p53 function.[52-54] It is thus proposed that oncogenic events in somatic tissues in AT due to loss of ATM can provide insights into malignant transformation of human cells with apparently normal ATM protein by activation of endogenous transposons and infection with oncogenic viruses.

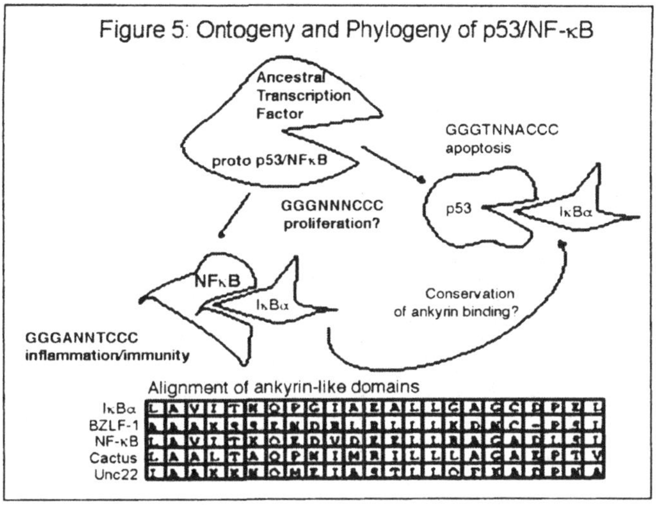

Figure 5. As shown in this figure (box at lower portion of the figure), the BZLF-1 protein encoded by EBV and expressed during viral lytic replication may encode an ankyrin-like region accounting for simultaneous interactions between p53 and NF-κB during EBV infection of lymphocytes. Notably other EBV encoded proteins expressed during viral latency alter both p53 and NF-κB expression and function during viral latency through a variety of mechanisms.

Evidence of Additional Specific Interactions between Epstein-Barr Virus (EBV) and V(D)J Recombination

The previous discussion has introduced the role of ATM as a protein required for "taming" of the wild V(D)J recombinase and other V(D)J like endogenous elements through detection of aberrant V(D)J recombination products at the immunoglobulin and T-cell receptor locus and also at endogenous mobile elements of the Tc family and at random genomic sites resembling V(D)J RSS. These effects may be amplified by the ability of oncogenic viral agents or cell stress to increase NF-κB transcription and thus block p53 effector functions, leading to a functional inactivity of ATM.

There is also evidence that oncogenic herpes viruses such as EBV may interact even more directly with the V(D)J recombination system in addition to their effects on the ATM and p53 pathways. Herpes viruses represent an episomal population present in most human somatic cells.[55] The oncogenic potential of some of these elements, particularly Epstein Barr Virus (EBV) is well established as well as the propensity for the EBV genome to insert into somatic genomes.

As shown in Figure 6, the terminal repeats of EBV[56,57] contain sequences with some similarity to V(D)J RSS as well as sequences with similar structure to the class switch signals.[58,59] Remarkably, the EBV genome contains a fragment of the entire IgG locus.[60] A variety of regulatory interactions are evident between EBV, V(D)J and class switch recombination and expression of the RAG genes, for example infection of lymphocytes activates RAG gene expression.[61,62] However, an EBV protein (the product of the BALF-2 ORF), postulated to be the viral recombinase binds to a Mg^+ required for its recombinase, functions and has a similar genomic structure and protein structure to the RAG-1 protein, a Mg binding DDE recombinase.[29]

Coincidentally, the replication of EBV genome and the genome of the Herpes viruses, synchronised with V(D)J recombination but not during the S-Phase of the cell cycle during which other homologous recombination events occur.[63-65] Coincidentally, the EBV BALF-2 protein expression is regulated by cAMP, Sp1, Ap1 and other signals that coregulate RAG expression and ligation

Sequences Resembling V(D)J RSS Adjacent To SP1/Class switch-like sites in EBV Terminal Repeats

GATCGGGG GGGTCGGGGG
GCGCCGCGCG GGCGCAGCCA TGCGTGACCG TGATGAGGGG GCAGGGTCGC
AGGGGGTGTG TCTGGTGGGG GGGGCGGCGC GGGAGCCTGC ACGCCGTTGG
AGGGTAGAAT GACAGGGGGC GGGGACAGAG AGGCGGTCGC GCCCCCGGCC
GCGCCAGCCA AGCCCCCAAG GGGGGCGGGG AGCGGGCAAT GGAGCGTGAC
GAAGGGCCCC AGGGCTGACC CCGGCAAACG TGACCCGGGG CTCCGGGGTG
ACCCAGCCAA GCGTGACCAA GGGGCCCGTG GGTGA**CACAG GCAACCCTGA**
CAAAGGCCCC CCAGGAAAGA CCCCC*GGGGG GCATCGGGGG GTGGGGCATG*
*GGGGG*CCGCG CATTCCTGGA AAAAGTGGAG GGGGCGTGGC CTTCCCCCGC
GGCCCCCCAG CCCCCCCGCA CAGAGCGGCG CTACGGCGGG CGGGCGGCGG
GGGGTCGGGG TCCGCGGGCT CCGGGGGCTG CGGGCGGTGG ATGGCGGCGG
ACGTTCCGGG **GATC**

Epstein-Barr TR repeat unit from SauIIIA sites (GATC)
VDJ-like sequences in bold
TRBP/Sp1 sites in gray

Figure 6. Showing functional V(D)J RSS and class switch elements in EBV replication as described.[29] A single copy of the viral terminal repeat bounded by SauIIIA restriction sites is shown. The EBV genome contains sequences with some similarity to V(D)J RSS and Tc element termini as noted previously (Fig. 1A,B), as well as class switch like G/C rich regions binding Sp1 transcription factors described in more detail elsewhere. These sequences are not randomly distributed in the virus, but are clustered in a repeated array termed the viral terminal repeat that undergoes replication dependent expansion and contraction as well as more complicated inversion and deletion events. In the case of other herpes viruses such as *H. simplex*, similar V(D)J like sequences are located in the regions of the *H. simplex* that undergoes site specific inversion during viral replication.

of the IgG receptor in B lymphocytes activates BALF-2 expression and also RAG expression in lymphocytes (Dreyfus, unpublished observations).

Remarkably, the herpes ICP-8 protein[66,67] (the *H. simplex* homologue of the EBV viral recombinase denoted BALF-2 discussed above) coprecipitates with proteins such as ku and DNApk from herpes virus that coprecipitate with the RAG proteins and elimination of ku markedly stimulates herpes replication.[68] Without too great an imagination it can be postulated that the EBV virus and other herpes viridae use a pathway similar to V(D)J recombination or Tc element transposition to form circular episomes to enter latency and then linearize to exit latency as proposed by the author more than a decade ago. Unfortunately, the consequence of these regulatory interactions between V(D)J recombination and oncogenic herpes viruses such as EBV may be that EBV infection simultaneously activates oncogenic V(D)J recombination pathogenesis and inactivates ATM through viral inactivation of p53.

These observations are directly relevant to human cancers in an increasingly large sphere of concern due to the possibility of viral "hit and run" carcinogenesis in which the virus transiently infects a lymphocyte or epithelial cell causing oncogenic DNA recombination with subsequent loss of the virus as the malignant transformation of the cell progresses. Most well known, literature concerns the oncogenic potential of Herpes viruses such as EBV in lymphatic cells in which the virus is often identified integrated into the malignant genome.[69-71] More recently, it has been reported that human maternal infection in the first trimester results in a 4 fold relative risk of leukemia in the off-spring, an observation not found for any other infectious agent, consistent with a hit-and-run mechanism in the fetus.[72] EBV infects human epithelial cells as well as lymphocytes, suggesting a role for the virus as a hit-and-run cofactor in breast cancer tissues.[73-75]

Summary: DDE Recombination Pathogenesis, or "Transposons Gone Wild"

In this work, the role of ATM protein has been reviewed with particular attention to the its role in the prevention of malignant transformation of cells through detection and elimination of and incorrect recombination products. The DDE class of recombinase generates hairpin structures during the process of DNA breakage and ligation that must subsequently be recognized by the host and repaired. V(D)J recombination pathogenes, due to the RAG DDE enzymes, is a subset of these oncogenic events. However these events occur even in animals deficient for the RAG proteins.

Thus the defects resulting from the loss of ATM should more properly be termed "DDE recombination pathogenesis" to indicate the expanding universe of DDE-encoding agents capable of causing oncogenic changes and cancer. Mechanistic questions have begun to be answered over the past decade due to detailed studies on the particular role of the ATM protein as a detector of hairpin and hairpin-like structures generated by DDE enzymes required for generation of the immune repertoire. A more novel and controversial proposal in this chapter is that the defects in ATM protein function may also play a role in aberrant recombination events associated with ubiquitous endogenous human repetitive DNA sequences related to mobile DNA sequences termed transposons and during infection of cells with chronic viral pathogens of the Herpes virus family such as EBV.

These observations could be important in stimulating new therapies for prevention of cancer in both patients with and without AT. An unfortunate consequence of the shared DDE mechanism between the RAG proteins (termed a "tamed" transposase) and other "untamed" transposases such as retroviral integrases and a putative Herpes virus recombinase is that mutations in ATM, or other functional forms of ATM inactivation can cause reversion of the "tamed" transposase to a less tamed or more "wild" phenotype. Could this unfortunate state be reversed and the transposase "retamed" through targeted inactivation of DDE recombination?

Acknowledgements

This work was in part supported in the past by NIH grants T32-GM07288 and T32-AI07365 to DHD and is based upon experimental studies conducted at the National Jewish Medical Research Center in the laboratory of and with the assistance and support of Dr. EW Gelfand and at the Albert Einstein College of Medicine in the laboratory of and with the assistance and support of Dr SW Emmons. The author is currently associated with the Clinical Faculty of the Yale School of Medicine and also founder of a biotechnology company Keren Pharmaceutical devoted to novel vaccines and genetic silencing therapies of viral and immunologic diseases. The author also acknowledges helpful discussion with Drs C Feschotte, JF Jones, C Kelleher, DG Schatz and L Ghoda.

References

1. Rotman G, Shiloh Y. ATM: from gene to function. Hum Mol Genet 1998; 7(10):1555-63.
2. Xu Y, Ashley T, Brainerd EE et al. Targeted disruption of ATM leads to growth retardation, chromosomal fragmentation during meiosis, immune defects and thymic lymphoma. Genes Dev 1996; 10(19):2411-2422.
3. Perkins EJ, Nair A, Cowley DO et al. Sensing of intermediates in V(D)J recombination by ATM. Genes Dev 2002; 16(2):159-164.
4. Giovannetti A, Mazzetta F, Caprini E et al. Skewed T-cell receptor repertoire, decreased thymic output and predominance of terminally differentiated T-cells in ataxia telangiectasia. Blood 2002; 100(12):4082-4089.
5. Matei IR, Gladdy RA, Nutter LM et al. ATM deficiency disrupts Tcra locus integrity and the maturation of CD4+CD8+ thymocytes. Blood 2007; 109(5):1887-1896.

6. Lantelme E, Turinetto V, Mantovani S et al. Analysis of secondary V(D)J rearrangements in mature, peripheral T-cells of ataxia telangiectasia heterozygotes. Lab Invest 2003; 83(10):1467-1475.
7. Lobrich M, Jeggo PA. Harmonising the response to DSBs: a new string in the ATM bow. DNA Repair (Amst) 2005; 4(7):749-759.
8. Revy P, Buck D, le Deist F et al. The repair of DNA damages/modifications during the maturation of the immune system: lessons from human primary immunodeficiency disorders and animal models. Adv Immunol 2005; 87:237-295.
9. Bredemeyer AL, Sharma GG, Huang CY et al. ATM stabilizes DNA double-strand-break complexes during V(D)J recombination. Nature 2006; 442(7101):466-470.
10. Ito K, Takubo K, Arai F et al. Regulation of reactive oxygen species by Atm is essential for proper response to DNA double-strand breaks in lymphocytes. J Immunol 2007; 178(1):103-110.
11. Barlow C, Brown KD, Deng CX et al. Atm selectively regulates distinct p53-dependent cell-cycle checkpoint and apoptotic pathways. Nat Genet 1997; 17(4):453-456.
12. Lu CP, Sandoval H, Brandt VL et al. Amino acid residues in Rag1 crucial for DNA hairpin formation. Nat Struct Mol Biol 2006; 13(11):1010-1015.
13. Raghavan SC, Kirsch IR et al. Analysis of the V(D)J recombination efficiency at lymphoid chromosomal translocation breakpoints. J Biol Chem 2001; 276(31):29126-29133.
14. Kitagawa Y, Inoue K, Sasaki S et al. Prevalent involvement of illegitimate V(D)J recombination in chromosome 9p21 deletions in lymphoid leukemia. J Biol Chem 2002; 277(48):46289-46297.
15. Lieber MR, Yu K, Raghavan SC. Roles of nonhomologous DNA end joining, V(D)J recombination and class switch recombination in chromosomal translocations. DNA Repair (Amst) 2006; 5(9-10):1234-1245.
16. Raghavan SC, Gu J, Swanson PC et al. The structure-specific nicking of small heteroduplexes by the RAG complex: Implications for lymphoid chromosomal translocations. DNA Repair (Amst) 2007; 6(6):751-759.
17. Fugmann SD, Lee AI, Shockett PE et al. The RAG proteins and V(D)J recombination: complexes, ends and transposition. Annu Rev Immunol 2000; 18:495-527.
18. Gellert M. V(D)J recombination: RAG proteins, repair factors and regulation. Annu Rev Biochem 2002; 71:101-132.
19. Wong SY, Roth DB. Murine models of Omenn syndrome. J Clin Invest 2007; 117(5):1213-1216.
20. Schwarz K, Gauss GH, Ludwig L et al. RAG mutations in human B-cell-negative SCID. Science 1996; 274(5284):97-99.
21. Jones JM, Gellert M. The taming of a transposon: V(D)J recombination and the immune system. Immunol Rev 2004; 200:233-248.
22. Shih IH, Melek M, Jayaratne ND et al. Inverse transposition by the RAG1 and RAG2 proteins: role reversal of donor and target DNA. EMBO J 2002; 21(23):6625-6633.
23. Oettinger MA. How to keep V(D)J recombination under control. Immunol Rev 2004; 200:165-181.
24. Chatterji M, Tsai CL, Schatz DG. Mobilization of RAG-generated signal ends by transposition and insertion in vivo. Mol Cell Biol 2006; 26(4):1558-1568.
25. Franco S, Alt FW, Manis JP. Pathways that suppress programmed DNA breaks from progressing to chromosomal breaks and translocations. DNA Repair (Amst) 2006; 5(9-10):1030-1041.
26. Liao MJ, Van Dyke T. Critical role for Atm in suppressing V(D)J recombination-driven thymic lymphoma. Genes Dev 1999; 13(10):1246-1250.
27. Tomimatsu N, Tahimic CG, Otsuki A et al. Ku70/80 Modulates ATM and ATR Signaling Pathways in Response to DNA Double Strand Breaks. J Biol Chem 2007; 282(14):10138-10145.
28. Rooney S, Alt FW, Sekiguchi J et al. Artemis-independent functions of DNA-dependent protein kinase in Ig heavy chain class switch recombination and development. Proc Natl Acad Sci USA 2005; 102(7):2471-2475.
29. Dreyfus DH. The DDE recombinases: diverse roles in acquired and innate immunity. Ann Allergy Asthma Immunol 2006; 97(5):567-76; quiz 76-8, 602.
30. Lee GS, Neiditch MB, Sinden RR et al. Targeted transposition by the V(D)J recombinase. Mol Cell Biol 2002; 22(7):2068-2077.
31. Neiditch MB, Lee GS, Huye LE et al. The V(D)J recombinase efficiently cleaves and transposes signal joints. Mol Cell 2002; 9(4):871-878.
32. Posey JE, Pytlos MJ, Sinden RR et al. Target DNA structure plays a critical role in RAG transposition. PLoS Biol 2006; 4(11):e350.
33. Tevelev A, Schatz DG. Intermolecular V(D)J recombination. J Biol Chem 2000; 275(12):8341-8348.
34. Tsai CL, Schatz DG. Regulation of RAG1/RAG2-mediated transposition by GTP and the C-terminal region of RAG2. EMBO J 2003; 22(8):1922-1930.
35. Chatterji M, Tsai CL, Schatz DG. New concepts in the regulation of an ancient reaction: transposition by RAG1/RAG2. Immunol Rev 2004; 200:261-271.

36. Kirsch IR, Lista F. Lymphocyte-specific genomic instability and risk of lymphoid malignancy. Semin Immunol 1997; 9(3):207-215.
37. Fuscoe JC, Vira LK, Collard DD et al. Quantification of hprt gene deletions mediated by illegitimate V(D)J recombination in peripheral blood cells of humans. Environ Mol Mutagen 1997; 29(1):28-35.
38. Meydan D, Nilsson T, Tornblom M et al. The frequency of illegitimate TCRbeta/gamma gene recombination in human lymphocytes: influence of age, environmental exposure and cytostatic treatment and correlation with frequencies of t(14;18) and hprt mutation. Mutat Res 1999; 444(2):393-403.
39. Gladdy RA, Taylor MD, Williams CJ et al. The RAG-1/2 endonuclease causes genomic instability and controls CNS complications of lymphoblastic leukemia in p53/Prkdc-deficient mice. Cancer Cell 2003; 3(1):37-50.
40. Dreyfus DH, Gelfand EW. Comparative analysis of invertebrate Tc6 sequences that resemble the vertebrate V(D)J recombination signal sequences (RSS). Mol Immunol 1999; 36(7):481-488.
41. Cordaux R, Udit S, Batzer MA et al. Birth of a chimeric primate gene by capture of the transposase gene from a mobile element. Proc Natl Acad Sci USA 2006; 103(21):8101-8106.
42. Feschotte C. Merlin, a new superfamily of DNA transposons identified in diverse animal genomes and related to bacterial IS1016 insertion sequences. Mol Biol Evol 2004; 21(9):1769-1780.
43. Liu D, Bischerour J, Siddique A et al. The human SETMAR protein preserves most of the activities of the ancestral Hsmar1 transposase. Mol Cell Biol 2007; 27(3):1125-1132.
44. Pace JK 2nd, Feschotte C. The evolutionary history of human DNA transposons: evidence for intense activity in the primate lineage. Genome Res 2007; 17(4):422-432.
45. Pritham EJ, Putliwala T, Feschotte C. Mavericks, a novel class of giant transposable elements widespread in eukaryotes and related to DNA viruses. Gene 2007; 390(1-2):3-17.
46. Wu A, Oh S, Ericson K et al. Transposon-based interferon gamma gene transfer overcomes limitations of episomal plasmid for immunogene therapy of glioblastoma. Cancer Gene Ther 2007; 14(6):550-560.
47. Piret B, Schoonbroodt S, Piette J. The ATM protein is required for sustained activation of NF-kappaB following DNA damage. Oncogene 1999; 18(13):2261-2271.
48. Reinhardt HC, Aslanian AS, Lees JA et al. p53-deficient cells rely on ATM- and ATR-mediated checkpoint signaling through the p38MAPK/MK2 pathway for survival after DNA damage. Cancer Cell 2007; 11(2):175-189.
49. Dreyfus DH, Nagasawa M, Gelfand EW et al. Modulation of p53 activity by IkappaBalpha: evidence suggesting a common phylogeny between NF-kappaB and p53 transcription factors. BMC Immunol 2005; 6:12.
50. Song H, Hollstein M, Xu Y. p53 gain-of-function cancer mutants induce genetic instability by inactivating ATM. Nat Cell Biol 2007; 9(5):573-580.
51. Suzuki K, Inageda K, Nishitai G et al. Phosphorylation of p53 at serine 15 in A549 pulmonary epithelial cells exposed to vanadate: Involvement of ATM pathway. Toxicol Appl Pharmacol 2007; 220(1):83-91.
52. Dreyfus DH, Nagasawa M, Kelleher CA et al. Stable expression of Epstein-Barr virus BZLF-1-encoded ZEBRA protein activates p53-dependent transcription in human Jurkat T-lymphoblastoid cells. Blood 2000; 96(2):625-634.
53. Mauser A, Saito S, Appella E et al. The Epstein-Barr virus immediate-early protein BZLF1 regulates p53 function through multiple mechanisms. J Virol 2002; 76(24):12503-12512.
54. Sinclair A, Yarranton S, Schelcher C. DNA-damage response pathways triggered by viral replication. Expert Rev Mol Med 2006; 8(5):1-11.
55. Kolman JL, Kolman CJ, Miller G. Marked variation in the size of genomic plasmids among members of a family of related Epstein-Barr viruses. Proc Natl Acad Sci USA 1992; 89(16):7772-7776.
56. Heller M, Henderson A, Kieff E. Repeat array in Epstein-Barr virus DNA is related to cell DNA sequences interspersed on human chromosomes. Proc Natl Acad Sci USA 1982; 79(19):5916-5920.
57. Raab-Traub N, Flynn K. The structure of the termini of the Epstein-Barr virus as a marker of clonal cellular proliferation. Cell 1986; 47(6):883-889.
58. Sun R, Spain TA, Lin SF et al. Autoantigenic proteins that bind recombinogenic sequences in Epstein-Barr virus and cellular DNA. Proc Natl Acad Sci USA 1994; 91(18):8646-8650.
59. Sun R, Spain TA, Lin SF et al. Sp1 binds to the precise locus of end processing within the terminal repeats of Epstein-Barr virus DNA. J Virol 1997; 71(8):6136-6143.
60. Niller HH, Salamon D, Rahmann S et al. A 30 kb region of the Epstein-Barr virus genome is colinear with the rearranged human immunoglobulin gene loci: implications for a "ping-pong evolution" model for persisting viruses and their hosts. A review. Acta Microbiol Immunol Hung 2004; 51(4):469-484.
61. Tsimbouri P, Drotar ME, Coy JL et al. bcl-xL and RAG genes are induced and the response to IL-2 enhanced in EmuEBNA-1 transgenic mouse lymphocytes. Oncogene 2002; 21(33):5182-5187.

62. Wagner HJ, Scott RS, Buchwald D et al. Peripheral blood lymphocytes express recombination-activating genes 1 and 2 during Epstein-Barr virus-induced infectious mononucleosis. J Infect Dis 2004; 190(5):979-984.
63. Cayrol C, Flemington E. G0/G1 growth arrest mediated by a region encompassing the basic leucine zipper (bZIP) domain of the Epstein-Barr virus transactivator Zta. J Biol Chem 1996; 271(50):31799-31802.
64. Altmann M, Hammerschmidt W. Epstein-Barr virus provides a new paradigm: a requirement for the immediate inhibition of apoptosis. PLoS Biol 2005; 3(12):e404.
65. Thorley-Lawson DA. EBV the prototypical human tumor virus—just how bad is it? J Allergy Clin Immunol 2005; 116(2):251-261; quiz 62.
66. Mapelli M, Panjikar S, Tucker PA. The crystal structure of the herpes simplex virus 1 ssDNA-binding protein suggests the structural basis for flexible, cooperative single-stranded DNA binding. J Biol Chem 2005; 280(4):2990-2997.
67. Uprichard SL, Knipe DM. Conformational changes in the herpes simplex virus ICP8 DNA-binding protein coincident with assembly in viral replication structures. J Virol 2003; 77(13):7467-7476.
68. Taylor TJ, Knipe DM. Proteomics of herpes simplex virus replication compartments: association of cellular DNA replication, repair, recombination and chromatin remodeling proteins with ICP8. J Virol 2004; 78(11):5856-5866.
69. Henderson A, Ripley S, Heller M et al. Chromosome site for Epstein-Barr virus DNA in a Burkitt tumor cell line and in lymphocytes growth-transformed in vitro. Proc Natl Acad Sci USA 1983; 80(7):1987-1991.
70. Okano M, Gross TG. A review of Epstein-Barr virus infection in patients with immunodeficiency disorders. Am J Med Sci 2000; 319(6):392-396.
71. Ambinder RF. Epstein-Barr virus-associated lymphoproliferative disorders. Rev Clin Exp Hematol 2003; 7(4):362-374.
72. Tedeschi R, Bloigu A, Ogmundsdottir HM et al. Activation of maternal Epstein-Barr virus infection and risk of acute leukemia in the offspring. Am J Epidemiol 2007; 165(2):134-137.
73. Huang J, Chen H, Hutt-Fletcher L et al. Lytic viral replication as a contributor to the detection of Epstein-Barr virus in breast cancer. J Virol 2003; 77(24):13267-13274.
74. Arbach H, Viglasky V, Lefeu F et al. Epstein-Barr virus (EBV) genome and expression in breast cancer tissue: effect of EBV infection of breast cancer cells on resistance to paclitaxel (Taxol). J Virol 2006; 80(2):845-853.
75. Perkins RS, Sahm K, Marando C et al. Analysis of Epstein-Barr virus reservoirs in paired blood and breast cancer primary biopsy specimens by real time PCR. Breast Cancer Res 2006; 8(6):R70.

Chapter 10

Ataxia Telangiectasia and Its Overlap with Nijmegen Breakage Syndrome and Ataxia-Like Disorders

Lindsay G. Ball and Wei Xiao*

Abstract

Ataxia Telangiectasia (AT), Nijmegen Breakage Syndrome (NBS) and AT-Like Disorder (ATLD) are three closely related human diseases. AT, NBS and ATLD share several prominent cellular phenotypes including increased sensitivity to ionizing radiation, abnormal cell-cycle checkpoints, chromosome instability, immunodeficiency and accelerated shortening of telomeres. However, notable distinctions amongst AT, NBS and ATLD patients are also prevalent. While AT patients are defective in the *ATM* gene, ATLD and NBS patients have mutations in genes encoding an MRN protein complex consisting of Mre11 (mutated in ATLD), Rad50 and Nbs1 (mutated in NBS). The MRN complex plays critical and complicated roles in regulating ATM functions which, upon examination, justifies the similarities and distinctions between AT, NBS and ATLD patients. Continued research and investigation into the mechanisms and pathways involved in MRN and ATM signaling will facilitate advanced diagnosis and treatment of AT and related diseases.

Introduction

There are several human deficiencies and diseases that are closely related to Ataxia Telangiectasia (AT). The establishment of AT as a monogenetic disease assisted in efficient diagnosis and reevaluation of "AT variants". Molecular cloning has allowed for the distinction between AT and other autosomal recessive cerebellar ataxias (ARCA) such as Friedreich ataxia or AT $_{Fresno}$, an old-fashioned term for an AT variant,[1-3] oculomotor apracias 1 (aprataxin deficiency), oculomotor apraxias 2 (senataxin deficiency), aicardi syndrome and AT-Like Disorder (ATLD).[4-7] Nijmegen Breakage Syndrome (NBS) was previously considered an AT variant, as several clinical features between AT and NBS overlap. As well, NBS and ATLD cells share a number of cellular phenotypes with AT cells, with the most prominent being the increased sensitivity to ionizing radiation (IR), abnormal cell-cycle checkpoints, chromosome instability, immunodeficiency and accelerated shortening of telomeres.[8,9] While all AT patients contain mutations in the AT-mutated (*ATM*) gene, most other ATLDs are defective in genes encoding an MRN protein complex consisting of Mre11 (mutated in ATLD), Rad50 and Nbs1 (mutated in NBS). Unlike AT, which can result from complete inactivation of the coding gene, all NBS and ATLD patients carry hypomorphic mutations that express some levels of corresponding protein, either truncated or full length with amino acid

*Corresponding Author: Wei Xiao—Department of Microbiology and Immunology, University of Saskatchewan, 107 Wiggins Road, Saskatoon, SK, S7N 5E5 Canada. Email: wei.xiao@usask.ca

Molecular Mechanisms of Ataxia Telangiectasia, edited by Shamim I. Ahmad.

substitutions.[5,10,11] The similarity between AT, NBS and ATLD signifies deficiency at common cellular function(s) and genetic pathways which will be further examined in this chapter.

Nijmegen Breakage Syndrome

Clinical Presentation of NBS

NBS is an autosomal recessive disease, which has also been referred to as chromosome instability syndrome and was first described in 1981 in Dutch patients with the majority of NBS patients living in the Czech Republic and Poland.[1,10] NBS patients have characteristic facial features or "bird-like" appearance which include: a receding forehead, receding mandible, prominent midface, large ears, epicanthal folds, sparse hair and microcephaly, which becomes apparent a few months after birth[9,10] (Table 1). Slow growth and a shortened body trunk resulting in a short stature are typically apparent by the age of two and are hallmarks of NBS patients. However, body weight is proportional to height for NBS patients.[12] NBS patients generally have an intelligence quotient (IQ) within normal range during early childhood; however, mental retardation progresses as they age. Dysgenesis with primary amenorrhea and elevated gonadotrophin levels have been described in several NBS patients.[13,14] A hallmark of NBS patients is their extreme sensitivity to IR, which is known to cause DNA double-strand breaks (DSBs). This sensitivity is characterized by an increase in spontaneous and IR induced chromosome breaks.[10,15,16] Cells from NBS patients do not slow down DNA replication or cell-cycle progression in the presence of DNA damaging agents such as bleomycin or IR, a dysfunction known as radioresistant DNA synthesis (RDS).[15-17] This has been demonstrated by an increase in chromosome aberrations following IR in cultured NBS cells when compared to normal cells, with the damage typically consisting of the chromatid-type.[10,16,18,19] NBS patients are also at high risk for developing cancer.[20] Children with NBS suffer from malignancies such as non-Hodgkin's lymphoma and B-cell lymphomas, which are rarely seen in children.[21] If NBS is undiagnosed before IR is used as a therapeutic agent (for example to treat Medulloblastoma) the result can be fatal.[10,22] NBS patients are also sensitive to other mutagens that cause DSBs such as bleomycin, streptonigrin, etoposide, camptothecin and mitomycin C.[10,23,24] NBS patients are prone to numerous types of infections, particularly respiratory tract infections, as a result of low levels of serum IgG subclasses and/or defective synthesis of specific antibodies to common pathogens.[25,26]

NBS and the NBS1 Gene

The gene involved in NBS phenotypes has been identified to be *NBS1* and mutations in the *NBS1* gene are responsible for NBS phenotypes. *NBS1* was identified in 1998 and is involved in the repair of DSBs.[3,27-29] The *NBS1* gene has been localized to chromosome 8q21 by linkage analysis in NBS families and identified by positional cloning.[3,28,30] The *NBS1* gene product is known as Nbs1, p95 or Nibrin. Nbs1 has 4 known functional regions: the N-terminal forkhead-associated (FHA) domain followed by a BRCA1 C-terminus (BRCT) domain, a C-terminal Mre11-binding domain and an ATM-binding domain at the extreme C-terminus[3,11] (Fig. 1A). Over 90% of all NBS patients analyzed to date contain a homozygous 5-bp truncation mutation, 657Δ5,[3] thus resulting in the production of a 26 kDa protein with FHA and BRCT domains but lacking the Mre11-binding domain. Additionally, there are seven mutations that have been identified in NBS patients. Knockout mice, producing similarly truncated Nbs1 proteins, are viable and develop symptoms characteristic of the human disease, allowing for a feasible animal model for NBS research.[31,32] Mice containing hypomorphic *NBS1* mutations survive; however, they demonstrate IR sensitivity, indicating the requirement for nibrin/Nbs1 phosphorylation by ATM as an early cellular response to IR.[31-33] In contrast, *NBS1*-null knockout mice are embryonically lethal.[34,35]

Nbs1 is the human ortholog to yeast Xrs2. The involvement of Xrs2 in nonhomologous end joining (NHEJ) suggests that the immunodeficiency of NBS patients could be attributed to a deficiency in V(D)J recombination.[10] In yeast, Xrs2 is part of a trimeric complex involving Mre11 and Rad50 (collectively termed MRX), which is known to be involved in DNA DSB repair, as is the mammalian Mre11, Rad50, Nbs1 (MRN) complex. The primary cellular function of

MRN is to sense DNA strand breaks and to activate signaling pathways leading to the cell-cycle checkpoint and recombination repair.[36-38] Thus, because Nbs1 and Mre11 are members of the same complex, their associated diseases NBS and ATLD are closely related.[1-3] NBS and ATLD have some overlapping clinical features such as hypersensitivity to IR and genome instability and both are characterized by neurological deficits,[36,39] indicating a deficiency at common cellular function(s) and genetic pathways.

AT-Like Disorder

Clinical Presentation of ATLD

ATLD is a very rare disorder with clinical features similar to those of AT and NBS, with the most prominent similarity being progressive cerebellar ataxia. However, ATLD patients do not present with telangiectasia[40,41] and thus have a later onset of neurological features, slower progression and milder symptoms compared to AT patients.[5] Clinical distinction made between NBS and ATLD is based on observations that NBS patients also present with characteristic facial features, microcephaly as well as growth retardation[42] (Table 1).

Table 1. Comparison of clinical features between AT, NBS and ATLD

Clinical Features	AT	NBS	ATLD
Abnormal eye movement	+	–	+
Abnormal cell-cycle checkpoints	+	+	+
Accelerated shortening of telomeres	+	+	+
Ataxia	+	–	+
Cancer predisposition	+	+	N/A
Chromosome instability	+	+	+
Chromosome translocation	+	+	+
Congenital malformation	–	+	–
Craniofacial abnormalities ("Bird-like" facial features)	–	+	–
Decreased homologous recombination	+	+	N/A
Dysarthria	+	–	+
Growth abnormalities	–	+	–
Hypersensitivity to IR	+	+	+
Increased AFP levels	+	–	–*
Microcephaly	–	+	–
Neurological features	+	+**	–
Neuronal degeneration	+	–	+
Normal intelligence	+	+/–	+
Ovarian failure	+	+	N/A
Pulmonary infections	+	+	N/A
Reduced immunoglobulin levels	+	+	–***
Skin abnormalities	+	+	N/A
Telangiectasia	+	–	–
Tumors	+	+	N/A

+: Presence of feature.

–: Absence of feature.

N/A: Information not available as too few patients have been described thus far.

*: Subject to debate.

**: NBS shows less extreme neurological features when compared with AT.

***: Deficiency in specific antibodies.

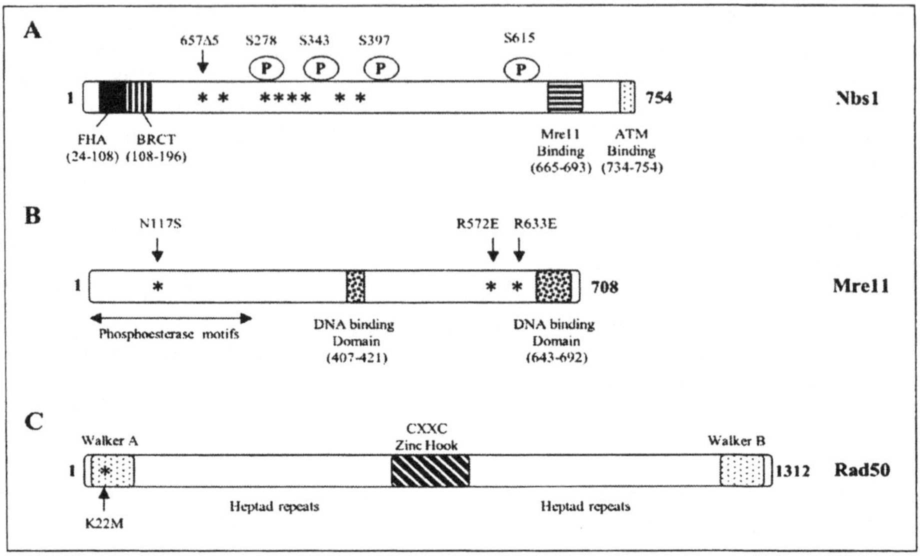

Figure 1. Schematic representation of the Nbs1, Mre11 and Rad50 proteins. A) Nbs1 and its functional domains. Asterisks represent mutations known to cause Nijmegen Breakage Syndrome (NBS) and the 657Δ5 mutation accounts for >90% of NBS patients. The letter P indicates serine locations of Nbs1 that are phosphorylated by ATM. B) Mre11 and its functional domains. ATLD is caused by mutations represented by the asterisks. C) Rad50 and its domain structure. Walker A and Walker B domains confer ATPase activity and are brought together via the heptad repeats and the resulting CXXC zinc hook aids interaction between Rad50 molecules. The K22M mutation is as noted in the *Rad50$^{S/S}$* mouse model.[66]

ATLD and the MRE11 Gene

ATLD is caused by mutations in the *MRE11* gene.[7] *MRE11* was initially isolated by its possible protein interaction with DNA ligase I and was predicted to encode a 708-amino acid protein.[43] However, subsequent reports favor an Mre11B sequence, differing from Mre11 primarily at the C-terminal region with significantly enhanced similarity to Mre11 from other species.[44] Mre11 contains a phosphoesterase motif at the N-terminal region that is highly conserved among all eukaryotic Mre11 and a similar motif is also found in *Escherichia coli* SbcD.[45] Mre11 also contains a DNA-binding domain at the C-terminal region and region(s) for homodimerization[46,47] (Fig. 1B). Attempts to create null *Mre11* mutations in mouse embryonic stem cells have failed[48] and conditional knockout experiments demonstrated that this gene is required for normal cell proliferation.[48] In mammalian cells, Mre11 also appears to be required for ATM activation[49] and an in vitro experiment showed that the MRN complex acts as a DSB sensor for ATM and recruits ATM to broken DNA molecules.[50] The unwinding of DNA ends by MRN appears to be essential for ATM stimulation implying that the MRN enzymatic activity plays an important role in this process.[50] The notion that MRN functions upstream of ATM is suggested by an observation that the extreme C-terminal 21 amino acid peptide of Nbs1 is required and sufficient for the interaction with ATM in vivo and in vitro and that this interaction mediates ATM deposition to the sites of DNA damage and its checkpoint functions.[51] These interactions between MRN and ATM begin to lay a foundation for understanding the molecular basis for the shared cellular functions in the cell-cycle checkpoints and recombination repair of DSB, as well as common syndromes among AT, NBS and ATLD.

The *RAD50* Gene

Mre11 binds directly to Nbs1, DNA and Rad50. In yeast the central region of Rad50 is essential for the activity of the MRX complex.[52-54] The numerous functions of the Mre11 subunit of the MRN complex are regulated through interactions with Rad50 and Nbs1.[29,36,55-57] The core Mre11-Rad50 (MR) complex, which is highly conserved, exists as a heterotetrameric complex (M_2R_2) which can be divided into 3 modular domains designated the head, coil and hook domain.[58,59] Rad50 in turn functions as a structural maintenance of chromosome (SMC) related protein employing its ATP-binding cassette (ABC) ATPase, Zn hook and coiled coils in order to bridge DSBs and facilitate DNA end processing by Mre11.[59,60] MRN and MRX DNA tethering, which has been observed microscopically and biochemically in vitro, has also been noted for *Saccharomyces cerevisiae* in vivo.[53,54,61] Initial observations suggested that conformational rotations in Rad50 ABC ATPase did not affect the overall MRN conformation and DNA tethering.[62] However, recent results demonstrate that the MRN complex contains both Rad50 ATPase and adenylate kinase catalytic activities, suggesting that the Rad50 nucleotide bound state mediates DNA bridging reactions.[63,64] Walker A and Walker B domains confer ATPase activity and are brought together via the heptad repeats and the resulting CXXC zinc hook supports the interaction between Rad50 molecules[36,65] (Fig. 1C). In *S. cerevisiae*, mutation of the Rad50 zinc hook, or deletion of any MRX component, disrupts end tethering. Unlike the *Rad50*$^{\Delta/\Delta}$ mouse, which is embryonically lethal, the *Rad50*$^{S/S}$ (hypomorphic *Rad50* mutant) mouse exhibits a shortened lifespan. *Rad50*$^{S/S}$ (or the *Rad50*^{K22M} allele which constitutes the mechanism for partial lethality in *Rad50*$^{S/S}$) mice exhibit total hematopoietic failure as a consequence of the gradual reduction in stem cells, growth defects and cancer predisposition.[66] With the tight interplay between members of the MRN complex and the underlying mechanisms by which NBS and ATLD share numerous clinical features with AT, it would thus not be surprising to identify patients with mutations in the *RAD50* gene. Therefore, continued research into understanding the underlying mechanisms of interplay amongst the MRN complex subunits and with ATM is vital to facilitate diagnosis and treatment of AT related diseases.

Underlying Mechanisms of the Interplay between AT, NBS and ATLD

The phenotypic overlap between AT, NBS and ATLD patients can be attributed to interrelated functions of their corresponding genes, *ATM*, *NBS1* and *MRE11*, respectfully. Although the exact mechanism of ATM activation has not been established, it is known to be heavily dependent on the MRN complex and subsequent posttranslational processes. Nevertheless, it should be noted that while MRN significantly enhances the activation of ATM, MRN is not essential for ATM's activation.[67] Mre11 and Rad50 sequences are highly conserved in eukaryotes, from unicellular yeasts to human.[68,69] The third component, Nbs1, is less conserved structurally among eukaryotes, although the mammalian Nbs1 and yeast Xrs2 play similar roles within their respective MRN/MRX complexes.

Following exposure to IR or radiomimetic agents, which typically results in DSBs, the MRN complex recognizes the damage and recruits ATM to bind to free DNA ends.[70-72] Not only does the MRN complex function upstream of ATM, but also it functions downstream of ATM in response to DSBs, a process described as a "two-way functional interaction".[73]

The initial function of the MRN complex in the "two-way functional interaction" is suggested to operate as a sensor for DNA DSBs. Following exposure to IR, MRN functions upstream of ATM, as demonstrated via the examination of a rationale that if MRN functioned upstream of ATM, ATM activation would be defective in cells from NBS and ATLD patients, as was found to be the case.[7,29,49] MRN rapidly locates to the site of DSBs[74,75] and the C-terminal portion of Nbs1 is imperative for the recruitment of ATM to the site of damage.[51] Mre11 interacts with both Rad50 and Nbs1 in the MRN complex and its homodimerization appears to be important for the Mre11-Rad50 and Mre11-Nbs1 interactions.[46,47] Mre11 deficiency results in decreased cellular levels of Rad50 and Nbs1, signifying that the MRN complex is crucial for maintaining protein stability. Mre11 does have endonuclease activity proven to be vital for ATM activation.[49]

In response to minimal IR (0.5 Gy), dimeric ATM (inactive) is also known to be autophosphorylated on Ser1981 (contained within the FAT domain of ATM), thus activating the monomeric ATM kinase.[76] Expression of the adenovirus proteins E1b55k/E4orf6, which degrades the MRN complex, abrogates ATM autophosphorylation as well as the ATM-dependent G2/M checkpoint,[77] again placing the MRN complex upstream of ATM and supporting the existence of a downstream function of MRN as well. Additionally, ATM has two other biologically significant autophosphorylation sites; S367 and S1893.[78,79] PP5 and Wip1 phosphatases are able to remove phosphates from S387 and S1981 in vitro and deficiency in either of these enzymes results in the upregulation of ATM kinase activity.[80-82]

IR is known to cause DNA DSBs, which lead to chromatin relaxation and altered DNA structures, resulting in the phosphorylation of H2AX.[83] This process is not required for ATM's function; however, phosphorylated H2AX may serve as scaffolding for subsequent DNA repair proteins. The Mre11-Rad50 complex is responsible for the stimulation of ATM, while Nbs1 is thought to be important for the translocation of Mre11-Rad50 to the nucleus and recruitment of ATM to the site of damage.[51,84]

After activation of ATM via the MRN complex and autophosphorylation, the kinase active ATM ensures subsequent pathways leading to cell-cycle checkpoint and recombination repair are initiated. ATM is also required to phosphorylate Mre11 following exposure to IR; however, the rationale behind this remains elusive.[85,86] ATM kinase is also required for phosphorylation of Nbs1 on Ser343 and 3 additional serine residues in response to DNA damage. Amino acid substitutions in the FHA and/or BRCT domains in Nbs1 result in the absence of phosphorylation at these serine residues.[87] The same amino acid substitutions that prevent phosphorylation of Nbs1 by ATM also prevent the accumulation of the MRN complex in nuclear foci.[87] The FHA and BRCT domains of Nbs1 are also required for interaction with histone H2AX, which is phosphorylated within minutes of exposure to IR at the site of DSBs.[88] These observations suggest that the interaction of H2AX with the Nbs1 FHA and BRCT domains is a prerequisite for efficient phosphorylation by ATM. ATM dependent phosphorylation of Nbs1 is important for activation of the S-phase checkpoint[33,89] and for cell cycle control in response to DSBs.[90,91] Nbs1 has thus been suggested to be a regulatory subunit of MRN that is essential for Mre11 phosphorylation upon DNA damage as well as Mre11's biochemical activities, such as ATP-dependent DNA unwinding and nuclease activity.[85,92] Neither Nbs1 nor Xrs2 is required for the enzymatic activity of Mre11-Rad50 in vitro; however, its activity appears to be absolutely required for MRX activity in vivo, as inactivation of any one of the three genes in *S. cerevisiae* cells leads to complete loss of MRX activity and indistinguishable cellular phenotypes.[11,93-95] The Nbs1 mutant S343A, which cannot be phosphorylated by ATM, is still capable of forming a complex with MRN, albeit no longer capable of activating the ATM kinase activity.[73,96]

ATM has been demonstrated to play a key role in all cell-cycle checkpoints[15,97-99] and its function with the S-phase checkpoint involves interaction with replication protein A (RPA).[100] ATM also phosphorylates p53 on Ser15, thus activating its transcriptional activity. Other ATM phosphorylation substrates include those involved in cell-cycle control, DNA repair and apoptosis.[101,102] Therefore, it is not surprising to observe pleotropic phenotypes of AT and its related disorders NBS and ATLD.

Conclusions

The physical and genetic interactions between the MRN complex and ATM play a central role in regulating ATM functions. These interactions appear to be complex and fit into a "two-way functional interaction" model, in which the MRN complex is required for enhancement of ATM activation as well as acting as the ATM kinase substrate. Unlike *ATM*, all three genes encoding the MRN complex are essential in mammals and mutations in *NBS1* and *MRE11*, that lose certain functions, result in NBS and ATLD respectively. The tight interplay between MRN and ATM provides insight into the underlying mechanisms by which NBS and ATLD share numerous clinical features with AT, whereas the wide range of phenotypes of NBS and ATLD patients reflects the hypomorphic nature of different mutations. In the future, it would not be surprising to identify patients with mutations in the *RAD50* gene, which would add new members to this class of diseases. Continual research to understand the mechanisms and pathways involved in MRN and ATM signaling is expected to improve the diagnosis and treatment of AT and AT related diseases.

Acknowledgements

The authors wish to thank Michelle Hanna for proofreading the manuscript and other laboratory members for useful discussion. This work was supported by the Canadian Institutes of Health Research operating grant MOP-53240 to WX.

References

1. Weemaes CM, Hustinx TW, Scheres JM et al. A new chromosomal instability disorder: the Nijmegen breakage syndrome. Acta Paediatr Scand 1981; 70:557-564.
2. Heil JA, Weemaes CM, van den Heuvel LP et al. International Nijmegen Breakage Syndrome Study Group: nijmegen breakage syndrome. Arch Dis Child 2000; 82:400-406.
3. Varon R, Vissinga C, Platzer M et al. Nibrin, a novel DNA double-strand break repair protein, is mutated in Nijmegen breakage syndrome. Cell 1998; 93:467-476.
4. Moreira MC, Klur S, Watanabe M et al. Senataxin, the ortholog of a yeast RNA helicase, is mutant in ataxia-ocular apraxia 2. Nat Genet 2004; 36:225-227.
5. Taylor AM, Groom A, Byrd PJ. Ataxia telangiectasia-like disorder (ATLD)-its clinical presentation and molecular basis. DNA Repair (Amst) 2004; 3:1219-1225.
6. Date H, Onodera O, Tanaka H et al. Early-onset ataxia with ocular motor apraxia and hypoalbuminemia is caused by mutations in a new HIT superfamily gene. Nat Genet 2001; 29:184-188.
7. Stewart GS, Maser RS, Stankovic T et al. The DNA double-strand break repair gene hMRE11 is mutated in individuals with an ataxia telangiectasia-like disorder. Cell 1999; 99:577-587.
8. Kobayashi J, Antoccia A, Tauchi H et al. NBS1 and its functional role in the DNA damage response. DNA Repair (Amst) 2004; 3:855-861.
9. Tauchi H, Matsuura S, Kobayashi J et al. Nijmegen breakage syndrome gene, NBS1 and molecular links to factors for genome stability. Oncogene 2002; 21:8967-8980.
10. Digweed M, Sperling K. Nijmegen breakage syndrome: clinical manifestation of defective response to DNA double-strand breaks. DNA Repair (Amst) 2004; 3:1207-1217.
11. Ball LG, Xiao W. Molecular basis of ataxia telangiectasia and related diseases. Acta Pharmacol Sin 2005; 26:897-907.
12. Chrzanowska KH, Kleijer WJ, Krajewska-Walasek M et al. Eleven Polish patients with microcephaly, immunodeficiency and chromosomal instability: the Nijmegen breakage syndrome. Am J Med Genet 1995; 57:462-471.
13. Chrzanowska KH, Krajewska-Walasek M, Bialecka M et al. Ovarian failure in female patients with the Nijmegen breakage syndrome. Eur J Hum Genet 1996; 4:122.
14. Conley ME, Spinner NB, Emanuel BS et al. A chromosomal breakage syndrome with profound immunodeficiency. Blood 1986; 67:1251-1256.
15. Painter RB, Young BR. Radiosensitivity in ataxia telangiectasia: a new explanation. Proc Natl Acad Sci USA 1980; 77:7315-7317.
16. Taalman RD, Jaspers NG, Scheres JM et al. Hypersensitivity to ionizing radiation, in vitro, in a new chromosomal breakage disorder, the Nijmegen Breakage Syndrome. Mutat Res 1983; 112:23-32.
17. D'Amours D, Jackson SP. The yeast Xrs2 complex functions in S-Phase checkpoint regulation. Genes Dev 2001; 15:2238-2249.
18. Antoccia A, Ricordy R, Maraschio P et al. Chromosomal sensitivity to clastogenic agents and cell cycle perturbations in Nijmegen breakage syndrome lymphoblastoid cell lines. Int J Radiat Biol 1997; 71:41-49.
19. Antoccia A, Stumm M, Saar K et al. Impaired p53-mediated DNA damage response, cell-cycle disturbance and chromosome aberrations in Nijmegen breakage syndrome lymphoblastoid cell lines. Int J Radiat Biol 1999; 75:583-591.
20. Seemanova E. An increased risk for malignant neoplasms in heterozygotes for a syndrome of microcephaly, normal intelligence, growth retardation, remarkable facies, immunodeficiency and chromosomal instability. Mutat Res 1990; 238:321-324.
21. Seidemann K, Tiemann M, Henze G et al. Therapy for non-Hodgkin's lymphoma in children with primary immunodeficiency: analysis of 19 patients from the BFM trials. Med Pediatr Oncol 1999; 33:536-544.
22. Distel L, Neubauer S, Varon R et al. Fatal toxicity following radio- and chemotherapy of medulloblastoma in a child with unrecognized Nijmegen breakage syndrome. Med Pediatr Oncol 2003; 41:44-48.

23. Nove J, Little JB, Mayer PJ et al. Hypersensitivity of cells from a new chromosomal-breakage syndrome to DNA-damaging agents. Mutat Res 1986; 163:255-62.
24. Kraakman-van der Zwet M, Overkamp WJ, Friedl AA et al. Immortalization and characterization of Nijmegen Breakage syndrome fibroblasts. Mutat Res 1999; 434:17-27.
25. Weemaes CM, Smeets DF, van der Burgt CJ. Nijmegen Breakage syndrome: a progress report. Int J Radiat Biol 1994; 66:S185-188.
26. Gregorek H, Chrzanowska KH, Michalkiewicz J et al. Heterogeneity of humoral immune abnormalities in children with Nijmegen breakage syndrome: an 8-year follow-up study in a single centre. Clin Exp Immunol 2002; 130:319-324.
27. van der Burgt I, Chrzanowska KH, Smeets D et al. Nijmegen breakage syndrome. J Med Genet 1996; 33:153-156.
28. Matsuura S, Tauchi H, Nakamura A et al. Positional cloning of the gene for Nijmegen breakage syndrome. Nat Genet 1998; 19:179-181.
29. Carney JP, Maser RS, Olivares H et al. The hMre11/hRad50 protein complex and Nijmegen breakage syndrome: linkage of double-strand break repair to the cellular DNA damage response. Cell 1998; 93:477-486.
30. Saar K, Chrzanowska KH, Stumm M et al. The gene for the ataxia telangiectasia variant, Nijmegen breakage syndrome, maps to a 1-cM interval on chromosome 8q21. Am J Hum Genet 1997; 60:605-610.
31. Williams BR, Mirzoeva OK, Morgan WF et al. A murine model of Nijmegen breakage syndrome. Curr Biol 2002; 12:648-653.
32. Kang J, Bronson RT, Xu Y. Targeted disruption of NBS1 reveals its roles in mouse development and DNA repair. EMBO J 2002; 21:1447-1455.
33. Gatei M, Young D, Cerosaletti KM et al. ATM-dependent phosphorylation of nibrin in response to radiation exposure. Nat Genet 2000; 25:115-119.
34. Dumon-Jones V, Frappart PO, Tong WM et al. Nbn heterozygosity renders mice susceptible to tumor formation and ionizing radiation-induced tumorigenesis. Cancer Res 2003; 63:7263-7269.
35. Zhu J, Petersen S, Tessarollo L et al. Targeted disruption of the Nijmegen breakage syndrome gene NBS1 leads to early embryonic lethality in mice. Curr Biol 2001; 11:105-109.
36. D'Amours D, Jackson SP. The Mre11 complex: at the crossroads of dna repair and checkpoint signalling. Nat Rev Mol Cell Biol 2002; 3:317-327.
37. Haber JE. The many interfaces of Mre11. Cell 1998; 95:583-586.
38. Symington LS. Role of RAD52 epistasis group genes in homologous recombination and double-strand break repair. Microbiol Mol Biol Rev 2002; 66:630-670, table of contents.
39. Petrini JH. The Mre11 complex and ATM: collaborating to navigate S-Phase. Curr Opin Cell Biol 2000; 12:293-296.
40. Klein C, Wenning GK, Quinn NP et al. Ataxia without telangiectasia masquerading as benign hereditary chorea. Mov Disord 1996; 11:217-220.
41. Hernandez D, McConville CM, Stacey M et al. A family showing no evidence of linkage between the ataxia telangiectasia gene and chromosome 11q22-23. J Med Genet 1993; 30:135-140.
42. Tauchi H, Kobayashi J, Morishima K et al. Nbs1 is essential for DNA repair by homologous recombination in higher vertebrate cells. Nature 2002; 420:93-98.
43. Petrini JH, Walsh ME, DiMare C et al. Isolation and characterization of the human MRE11 homologue. Genomics 1995; 29:80-86.
44. Chamankhah M, Wei YF, Xiao W. Isolation of hMRE11B: failure to complement yeast mre11 defects due to species-specific protein interactions. Gene 1998; 225:107-116.
45. Sharples GJ, Leach DR. Structural and functional similarities between the SbcCD proteins of Escherichia coli and the RAD50 and MRE11 (RAD32) recombination and repair proteins of yeast. Mol Microbiol 1995; 17:1215-1217.
46. Usui T, Ohta T, Oshiumi H et al. Complex formation and functional versatility of Mre11 of budding yeast in recombination. Cell 1998; 95:705-716.
47. Chamankhah M, Xiao W. Formation of the yeast Mre11-Rad50-Xrs2 complex is correlated with DNA repair and telomere maintenance. Nucleic Acids Res 1999; 27:2072-2079.
48. Xiao Y, Weaver DT. Conditional gene targeted deletion by Cre recombinase demonstrates the requirement for the double-strand break repair Mre11 protein in murine embryonic stem cells. Nucleic Acids Res 1997; 25:2985-2991.
49. Uziel T, Lerenthal Y, Moyal L et al. Requirement of the MRN complex for ATM activation by DNA damage. EMBO J 2003; 22:5612-5621.
50. Lee JH, Paull TT. ATM activation by DNA double-strand breaks through the Mre11-Rad50-Nbs1 complex. Science 2005; 308:551-554.
51. Falck J, Coates J, Jackson SP. Conserved modes of recruitment of ATM, ATR and DNA-PKcs to sites of DNA damage. Nature 2005; 434:605-611.

52. Hopfner KP, Craig L, Moncalian G et al. The Rad50 zinc-hook is a structure joining Mre11 complexes in DNA recombination and repair. Nature 2002; 418:562-566.
53. Lobachev K, Vitriol E, Stemple J et al. Chromosome fragmentation after induction of a double-strand break is an active process prevented by the RMX repair complex. Curr Biol 2004; 14:2107-2112.
54. Wiltzius JJ, Hohl M, Fleming JC et al. The Rad50 hook domain is a critical determinant of Mre11 complex functions. Nat Struct Mol Biol 2005; 12:403-407.
55. Moncalian G, Lengsfeld B, Bhaskara V et al. The rad50 signature motif: essential to ATP binding and biological function. J Mol Biol 2004; 335:937-951.
56. Paull TT, Gellert M. Nbs1 potentiates ATP-driven DNA unwinding and endonuclease cleavage by the Mre11/Rad50 complex. Genes Dev 1999; 13:1276-1288.
57. Paull TT, Gellert M. The 3' to 5' exonuclease activity of Mre 11 facilitates repair of DNA double-strand breaks. Mol Cell 1998; 1:969-979.
58. Hopfner KP, Karcher A, Craig L et al. Structural biochemistry and interaction architecture of the DNA double-strand break repair Mre11 nuclease and Rad50-ATPase. Cell 2001; 105:473-485.
59. Williams RS, Williams JS, Tainer JA. Mre11-Rad50-Nbs1 is a keystone complex connecting DNA repair machinery, double-strand break signaling and the chromatin template. Biochem Cell Biol 2007; 85:509-520.
60. Stracker TH, Theunissen JW, Morales M et al. The Mre11 complex and the metabolism of chromosome breaks: the importance of communicating and holding things together. DNA Repair (Amst) 2004; 3:845-854.
61. Kaye JA, Melo JA, Cheung SK et al. DNA breaks promote genomic instability by impeding proper chromosome segregation. Curr Biol 2004; 14:2096-2106.
62. Moreno-Herrero F, de Jager M, Dekker NH et al. Mesoscale conformational changes in the DNA-repair complex Rad50/Mre11/Nbs1 upon binding DNA. Nature 2005; 437:440-443.
63. Williams RS, Tainer JA. Learning our ABCs: Rad50 directs MRN repair functions via adenylate kinase activity from the conserved ATP binding cassette. Mol Cell 2007; 25:789-791.
64. Bhaskara V, Dupre A, Lengsfeld B et al. Rad50 adenylate kinase activity regulates DNA tethering by Mre11/Rad50 complexes. Mol Cell 2007; 25:647-661.
65. de Jager M, van Noort J, van Gent DC et al. Human Rad50/Mre11 is a flexible complex that can tether DNA ends. Mol Cell 2001; 8:1129-1135.
66. Bender CF, Sikes ML, Sullivan R et al. Cancer predisposition and hematopoietic failure in Rad50(S/S) mice. Genes Dev 2002; 16:2237-2251.
67. Horejsi Z, Falck J, Bakkenist CJ et al. Distinct functional domains of Nbs1 modulate the timing and magnitude of ATM activation after low doses of ionizing radiation. Oncogene 2004; 23:3122-3127.
68. Anderson RA, Boronenkov IV, Doughman SD et al. Phosphatidylinositol phosphate kinases, a multifaceted family of signaling enzymes. J Biol Chem 1999; 274:9907-9910.
69. Kunz J, Wilson MP, Kisseleva M et al. The activation loop of phosphatidylinositol phosphate kinases determines signaling specificity. Mol Cell 2000; 5:1-11.
70. Kurz EU, Lees-Miller SP. DNA damage-induced activation of ATM and ATM-dependent signaling pathways. DNA Repair (Amst) 2004; 3:889-900.
71. Smith GC, Cary RB, Lakin ND et al. Purification and DNA binding properties of the ataxia telangiectasia gene product ATM. Proc Natl Acad Sci USA 1999; 96:11134-11139.
72. Suzuki K, Kodama S, Watanabe M. Recruitment of ATM protein to double strand DNA irradiated with ionizing radiation. J Biol Chem 1999; 274:25571-25255.
73. Lavin MF. The Mre11 complex and ATM: a two-way functional interaction in recognising and signaling DNA double strand breaks. DNA Repair (Amst) 2004; 3:1515-1520.
74. Maser RS, Monsen KJ, Nelms BE et al. hMre11 and hRad50 nuclear foci are induced during the normal cellular response to DNA double-strand breaks. Mol Cell Biol 1997; 17:6087-6096.
75. Mirzoeva OK, Petrini JH. DNA damage-dependent nuclear dynamics of the Mre11 complex. Mol Cell Biol 2001; 21:281-288.
76. Bakkenist CJ, Kastan MB. DNA damage activates ATM through intermolecular autophosphorylation and dimer dissociation. Nature 2003; 421:499-506.
77. Carson CT, Schwartz RA, Stracker TH et al. The Mre11 complex is required for ATM activation and the G2/M checkpoint. EMBO J 2003; 22:6610-6620.
78. Kozlov S, Gueven N, Keating K et al. ATP activates ataxia telangiectasia mutated (ATM) in vitro. Importance of autophosphorylation. J Biol Chem 2003; 278:9309-9317.
79. Kozlov SV, Graham ME, Peng C et al. Involvement of novel autophosphorylation sites in ATM activation. EMBO J 2006; 25:3504-3514.
80. Ali A, Zhang J, Bao S et al. Requirement of protein phosphatase 5 in DNA-damage-induced ATM activation. Genes Dev 2004; 18:249-254.

81. Lu X, Nguyen TA, Donehower LA. Reversal of the ATM/ATR-mediated DNA damage response by the oncogenic phosphatase PPM1D. Cell Cycle 2005; 4:1060-1064.
82. Shreeram S, Demidov ON, Hee WK et al. Wip1 phosphatase modulates ATM-dependent signaling pathways. Mol Cell 2006; 23:757-764.
83. Shroff R, Arbel-Eden A, Pilch D et al. Distribution and dynamics of chromatin modification induced by a defined DNA double-strand break. Curr Biol 2004; 14:1703-1711.
84. Cerosaletti K, Concannon P. Independent roles for nibrin and Mre11-Rad50 in the activation and function of Atm. J Biol Chem 2004; 279:38813-38819.
85. Dong Z, Zhong Q, Chen PL. The Nijmegen breakage syndrome protein is essential for Mre11 phosphorylation upon DNA damage. J Biol Chem 1999; 274:19513-19516.
86. Yuan SS, Chang HL, Hou MF et al. Neocarzinostatin induces Mre11 phosphorylation and focus formation through an ATM- and NBS1-dependent mechanism. Toxicology 2002; 177:123-130.
87. Cerosaletti KM, Concannon P. Nibrin forkhead-associated domain and breast cancer C-terminal domain are both required for nuclear focus formation and phosphorylation. J Biol Chem 2003; 278:21944-21951.
88. Kobayashi J, Tauchi H, Sakamoto S et al. NBS1 localizes to gamma-H2AX foci through interaction with the FHA/BRCT domain. Curr Biol 2002; 12:1846-1851.
89. Lim DS, Kim ST, Xu B et al. ATM phosphorylates p95/nbs1 in an S-phase checkpoint pathway. Nature 2000; 404:613-617.
90. Yazdi PT, Wang Y, Zhao S et al. SMC1 is a downstream effector in the ATM/NBS1 branch of the human S-phase checkpoint. Genes Dev 2002; 16:571-582.
91. Nakanishi K, Taniguchi T, Ranganathan V et al. Interaction of FANCD2 and NBS1 in the DNA damage response. Nat Cell Biol 2002; 4:913-920.
92. Steimle PA, Hoffert JD, Adey NB et al. Polyphosphoinositides inhibit the interaction of vinculin with actin filaments. J Biol Chem 1999; 274:18414-18420.
93. Ivanov EL, Sugawara N, White CI et al. Mutations in XRS2 and RAD50 delay but do not prevent mating-type switching in Saccharomyces cerevisiae. Mol Cell Biol 1994; 14:3414-3425.
94. Johzuka K, Ogawa H. Interaction of Mre11 and Rad50: two proteins required for DNA repair and meiosis-specific double-strand break formation in Saccharomyces cerevisiae. Genetics 1995; 139:1521-1532.
95. Chamankhah M, Fontanie T, Xiao W. The Saccharomyces cerevisiae mre11(ts) allele confers a separation of DNA repair and telomere maintenance functions. Genetics 2000; 155:569-576.
96. Lee JH, Paull TT. Direct activation of the ATM protein kinase by the Mre11/Rad50/Nbs1 complex. Science 2004; 304:93-96.
97. Shiloh Y. Ataxia telangiectasia and the Nijmegen breakage syndrome: related disorders but genes apart. Annu Rev Genet 1997; 31:635-662.
98. Beamish H, Lavin MF. Radiosensitivity in ataxia telangiectasia: anomalies in radiation-induced cell cycle delay. Int J Radiat Biol 1994; 65:175-184.
99. Houldsworth J, Lavin MF. Effect of ionizing radiation on DNA synthesis in ataxia telangiectasia cells. Nucleic Acids Res 1980; 8:3709-3720.
100. Olson E, Nievera CJ, Liu E et al. The Mre11 complex mediates the S-phase checkpoint through an interaction with replication protein A. Mol Cell Biol 2007; 27:6053-6067.
101. Lavin MF, Gueven N. The complexity of p53 stabilization and activation. Cell Death Differ 2006; 13:941-950.
102. Lavin MF, Kozlov S. DNA damage-induced signalling in ataxia telangiectasia and related syndromes. Radiother Oncol 2007; 83:231-237.

Chapter 11

Animal Models for Ataxia Telangiectasia

Ramune Reliene and Robert H. Schiestl*

Abstract

Atm deficient (*Atm*$^{-/-}$, *Atm*$^{y/y}$ and *Atm*-ΔSRI) mice closely mimic most of the phenotype of human ataxia telangiectasia (AT). This includes growth retardation, immunodeficiency, infertility, genomic instability, acute sensitivity to ionizing radiation and a striking predisposition to lymphoid malignancies. However, *Atm* deficient mice do not recapitulate cerebellar degeneration and show only a mild form of ataxia. Nevertheless, *Atm* deficient mice provided researchers the first opportunity to study AT in a controlled way in mammals and in multiple tissues. Studies with *Atm* deficient mice revealed most of our understanding of the role of oxidative stress in AT. Consequently, oxidative stress has been viewed as a potential mechanism contributing to the clinical manifestations of AT. Thus, *Atm* deficient mice have been used to study the effect of antioxidants, as a new therapeutic approach to treat AT. These studies showed that antioxidants N-Acetyl-L- Cysteine (NAC), EUK-189, tempol and CTMIO are able to correct some cellular and clinical phenotypes of AT. In particular, antioxidants were efficient in preventing lymphomagenesis. An ongoing clinical trial (Thomas Jefferson University Medical College) and future chemoprevention trials in AT patients, will provide more evidence as to whether antioxidant intake can slow down disease progression in AT.

Introduction

Mouse Models for AT

Soon after the discovery of the human *ATM* gene located on human chromosome 11[1,2] researchers have focused on the equivalent murine *Atm* gene located on mouse chromosome 9.[3] Genetic targeting of the murine *Atm* gene, that is highly homologous to the corresponding human counterpart,[3] allowed generation of AT mouse models, *Atm* deficient mice, which provided researchers with the first opportunity to study the disease in a controlled way in a mammal. Three *Atm*$^{-/-}$ mouse strains,[4-6] *Atm*$^{y/y}$ line [7] and *Atm*-ΔSRI strain[8] have been generated in different laboratories. Generation of genetically engineered mice was a tremendous success as they closely mimic human AT characteristics. Similar to the human AT phenotype, *Atm* deficient mice show growth retardation, immunodeficiency, infertility, genomic instability, acute sensitivity to ionizing radiation and a striking predisposition to thymic lymphomas.[4-9] However, AT mouse models do not recapitulate a prime feature of the human disease, namely cerebellar degeneration and show only a mild form of ataxia.[4,6,7,10]

Atm$^{-/-}$ and *Atm*$^{y/y}$ mice were generated by disrupting the *Atm* gene (truncation mutations) resulting in the absence of detectable Atm protein,[4-7] while *Atm*-ΔSRI mice were generated by

*Corresponding Author: Robert H. Schiestl—Dept. of Pathology, UCLA School of Medicine, 650 Charles E. Young Drive South, Los Angeles, CA 90024. Email: rschiestl@mednet.ucla.edu

Molecular Mechanisms of Ataxia Telangiectasia, edited by Shamim I. Ahmad.

introducing a nine nucleotide in-frame deletion in the *Atm* gene (missense mutation) resulting in the loss of three amino acid (SRI) residues.[8] *Atm*-ΔSRI mice express nearly full-length mutant Atm protein that has only a low level of residual kinase activity, which does not respond to radiation.[8] This mutant Atm protein has a dominant-negative activity upon its normal counterpart, which predisposes *Atm*-ΔSRI heterozygous mice to a variety of tumors, while $Atm^{+/-}$ mice do no show increased tumorigenesis compared to wild type mice.[11]

Clinical Features of Atm Deficient Mice

Cancer

Atm deficient mice are an especially valuable tool to study cancer predisposition, as most of them succumb to thymic lymphomas irrespective of the transgenic model or genetic background.[4-8] On the other hand, AT patients are characterized by a wider spectrum of tumors. At young age AT patients primarily develop leukemias and lymphomas of T-cells and less frequently of B-cell origin, while older patients also develop solid tumors (oral cavity, breast, stomach, pancreas, ovary and bladder).[9,12,13] Nevertheless, lymphoma manifestation in *Atm* deficient mice is probably the best reproduced clinical feature of AT.

Thymic lymphomas in *Atm* deficient mice arise from highly proliferating CD4/CD8 double positive T-cells.[4,6] In some studies, tumors are widely metastatic,[4,14] while in others, not.[6] Tumors from *Atm* deficient mice exhibit multiple chromosomal aberrations, most frequently unbalanced rearrangements in chromosomes 12, 14 and 15 resulting in net deletion or duplication of genetic material.[15] Certain chromosomal translocations in tumors from *Atm* deficient mice involve T-cell receptor (TCR) α/δ locus, others involve random loci.[15] Thus, lymphomagenesis in *Atm* deficient mice is associated with genomic instability.

Remarkably, the onset of lymphoma varies from one laboratory to another and even from one study to another within the same laboratory.[16] In some, especially early reports, all $Atm^{-/-}$ mice died between 2 and 5 month of age.[4-6,17] In other studies, 50% of $Atm^{-/-}$ mice survived for 5-7 month[8,18,19] or 10-12 month.[14,20] It appears that longer life spans in $Atm^{-/-}$ mice are attributable to breeding in specific pathogen free (SPF) conditions rather than genetic background.[8,16] $Atm^{y/y}$ mice that were bred in SPF facilities also lived significantly longer, i.e., 50% survived for 10 month or more.[7] *Atm*-ΔSRI homozygous mice live about 2-fold longer than $Atm^{-/-}$ mice, provided that survival analyses were completed in the same study.[8] In this study, all $Atm^{-/-}$ mice died by 10 month, while about half of *Atm*-ΔSRI homozygous mice were still alive. Interestingly, *Atm*-ΔSRI homozygous mice that died by 10 month of age developed lymphoma, while mice that died at later age developed a variety of different tumors including B-cell tumors, sarcomas and carcinomas.[8] Animal age may be a major factor determining the appearance of solid tumors in *Atm* deficient mice. In our study, where $Atm^{-/-}$ mice lived markedly longer than in other studies resulting in 50% survival at 12.5 month, tumor spectra ranged from lymphomas (total lymphoma incidence 76%) to ovarian cancers (33%) and bronchoalveolar adenomas (6%).[14] Solid tumors were predominantly observed in older mice (>17 month old). Solid tumors were also observed in a study where 40% of $Atm^{-/-}$ mice lived more than 18 months,[20] but not in studies where all $Atm^{-/-}$ mice died by 4-5 months of age.[4,6]

Immunological Defects

Lymphoid malignancies in *Atm* deficient mice are associated with defective immune system development. *Atm* deficient mice have smaller thymuses and fewer thymocytes than wild type mice.[4-8] In particular, they have significantly reduced numbers of mature CD8 and CD4 single positive T-cells. However, the absolute number of early T-lineage precursor, CD4/CD8 double negative thymocytes, is similar in $Atm^{-/-}$ and wild type mice indicating defective immune cell differentiation and maturation.[6] Developmental defects are also characteristic for B-cells. For example, precursor B-cells and immature B-cells are decreased in the bone marrow of *Atm* deficient mice.[6,7] The T-dependent immune response is impaired in $Atm^{-/-}$ mice, while the B-cell response is normal.[6] Serum concentrations of immunoglobulin (Ig) isotypes IgM, IgG, IgA are similar, however, secretion of IgG subclasses $IgG_{2\alpha}$, $IgG_{2\beta}$ and IgG_3 is reduced.[6] Immunological defects in *Atm* deficient mice are attributable to aberrant V(D)J recombination between TCR and Ig genes.[21,22] V(D)J recombination, a process requiring regulated DNA double strand break (DSB) repair, may contribute to the formation of DSBs that lead to deleterious chromosomal rearrangements, however, it is not essential in lymphomagenesis of *Atm* deficient mice. When V(D)J recombination is blocked by mutational inactivation of recombination activating genes, *Rag1* and/or *Rag2*, double mutant *Rag1*$^{-/-}$*Atm*$^{-/-}$ and *Rag2*$^{-/-}$*Atm*$^{-/-}$ mice still develop thymic lymphomas.[23]

Neurological Symptoms

Unlike AT patients, *Atm* deficient mice do not exhibit gross cerebellar degeneration or overt ataxia.[4-8,24] However, mild motor deficit was consistently found in stringent behavioral studies.[4,7,25,26] In particular, *Atm* deficient mice showed some deficit in balancing[4] and motor learning on an accelerating rotating rod.[7] Gait analysis showed that stride length asymmetry, indicative of ataxia, is increased but these results were inconsistent in different studies.[4,10,26] Interestingly, one line of *Atm*$^{-/-}$ mice[5] has been shown to develop a selective loss of nigrostriatal dopamine neurons with age[10,27] similar to Parkinson's disease. In addition, ectopic and abnormally differentiated Purkinje cells were found in the cerebellum of *Atm*$^{y/y}$ mice.[7] Furthermore, in vitro survival of cerebellar Purkinje cells from *Atm*$^{-/-}$ and *Atm*-ΔSRI homozygous mice was reduced compared to wild type mice.[28] Taken together, *Atm* deficient mice are characterized by subtle neurological symptoms but it does not reproduce very well the phenotype of human AT patients.

Cellular Phenotypes of Atm Deficient Mice

Genomic Instability

Atm deficient mice recapitulate a hallmark of the cellular phenotype of AT cells, namely genomic instability. Cells from *Atm* deficient mice display a wide range of genomic alterations. MEFs, isolated from *Atm*$^{-/-}$, mice exhibit 6.5-fold more chromosomal breaks per metaphase than *Atm* heterozygous and wild type mice.[5] It has been reported that telomeres are shortened in splenocytes, thymocytes and bone marrow cells of *Atm*$^{-/-}$ mice.[29] As expected, telomere shortening is correlated with high levels of chromosomal abnormalities including fragmentations, end-to-end fusions deletions and aneuploidy.[29] *Atm*$^{-/-}$ mice display an elevated frequency of intrachromosomal homologous recombination resulting in DNA deletions.[30]

Oxidative Stress

A great deal of our understanding of oxidative stress in AT stems from animal studies. Similarly to human AT patients,[31] the presence of oxidative stress in *Atm* deficient mice was demonstrated by elevated levels of oxidatively damaged lipids, proteins and DNA.[18,32,33] This may be a reflection of an elevated intracellular level of reactive oxygen species (ROS), as determined in the brain, thymus and MEFs of *Atm* deficient mice.[18,19,34] Several studies focused on the brain, especially cerebellum, as a target organ of oxidative stress. These studies found that *Atm* deficient mice exhibit altered levels of enzymes that respond to oxidative stress. For example, heme oxygenase,[32] thioredoxin and superoxide dismutase (SOD) activities are increased, while catalase activity[35] and pyridine nucleotides NAD+, NADH and NADP+[36] are decreased in the brains of *Atm*$^{-/-}$ mice. The level of thiol-containing amino acid cysteine, which confers antioxidant activity, is decreased in cerebellum and cerebrum of *Atm*$^{-/-}$ mice at the age of 2-month.[35] These findings correlate with higher level of ROS that are selectively elevated in Purkinje cells and nigrostriatal dopaminergic cells.[34]

Experimental Disease Prevention in AT

AT mouse models not only allow studying the disease in a variety of cells and tissues but, most importantly, it allows testing new therapeutic approaches, which is not possible in human AT patients. Recently, antioxidants gained attention as a potential therapeutic means to prevent the disease progression in AT, inspired by the facts that a) AT is characterized by persistent oxidative stress and b) oxidative stress is linked to both neurodegenerative diseases and cancer.[37] In response to this view, several studies were undertaken to test the effect of antioxidants in *Atm* deficient mice, which includes N-Acetyl-L-Cysteine (NAC),[14] EUK-189,[38] tempol[18] and CTMIO.[25]

NAC

NAC, a pro-drug of cysteine and glutathione (GSH), is a drug that is safe and without major side-effects and has been used in clinical settings for more than four decades.[39,40] NAC has been widely used for the treatment of respiratory diseases as a mucolytic agent,[41] for acetaminophen overdose, where it rescues from GSH depletion in the liver.[42] and is available as an over-the-counter dietary supplement. Multiple clinically relevant effects have been ascribed to NAC. For example, 1) NAC directly scavenges ROS, 2) can enhance the synthesis of GSH and 3) reduces inflammation.[43]

We have undertaken both short term and long-term studies to understand whether NAC can prevent genomic instability and/or cancer in *Atm* deficient mice.[14,33] $Atm^{-/-}$ mice received NAC-supplemented drinking water chronically from fertilization throughout their life. The major reason to start antioxidant administration as early as from fertilization was to protect against genome rearrangements that can arise during mouse development and lead to carcinogenesis later in life. In fact, we found that NAC protects against DNA deletions that occur during embryonic development.[33] In addition, NAC prevents oxidative DNA damage in $Atm^{-/-}$ mice.[33] It was therefore expected that NAC has a potential to reduce carcinogenesis in $Atm^{-/-}$ mice. A long-term survival study provided exciting results that have clinical implications. We found that NAC significantly increased the lifespan and reduced both the incidence and multiplicity of lymphoma in *Atm* deficient mice.[14] The mean survival of NAC treated mice was 17 month, while that of untreated mice was only 12.5 month ($p = 0.03$). Most (76.5%) untreated mice were diagnosed with lymphoma and tumors were found in multiple lymphoid and nonlymphoid tissues indicating the aggressive nature of the tumors (Fig. 1A and B). Remarkably, the incidence of lymphoma in NAC treated mice was reduced by 2-fold (37.5 versus 76.5 %, $p = 0.02$) (Fig. 1A). Furthermore, the multiplicity of tumors decreased from 4.6 to 2.8 tumors per mouse ($p = 0.038$). Lymphoma burden was similar in the thymus, spleen and liver, while in other organs, such as lymph nodes, lung, heart, kidney, pancreas, stomach, duodenum and adrenal glands there were fewer or no tumors in the NAC treatment group (Fig. 1B).

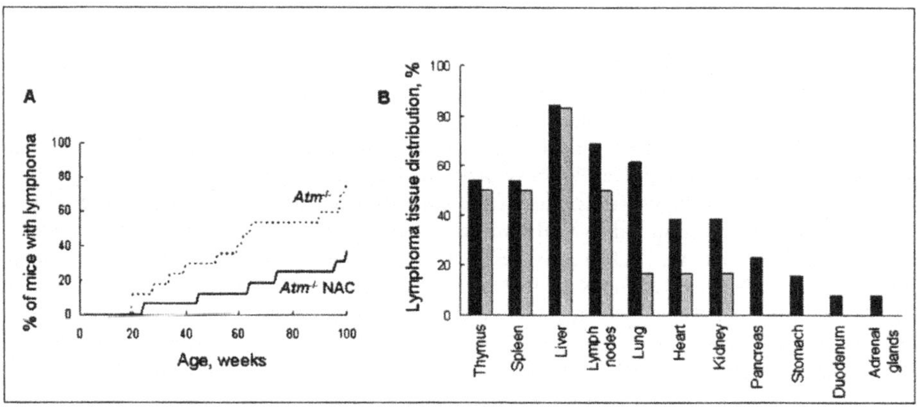

Figure 1. NAC reduced the incidence and multiplicity of lymphoma in *Atm* deficient mice. A) Percentage of mice with lymphoma is shown as a function survival time. B) The lymphoma tissue distribution. Only mice that had lymphoma are included in the calculation. Black bars depict untreated mice, gray bars show NAC treated mice. Lymph nodes affected were mesenteric and/or peripheral, thoratic and perirenal. Lymphoma in the heart was seen in epicardium and/or pericardium. Taken from Reliene R, Schiestl RH. DNA Repair (Amst) 2006; 5:852-859[14]; © 2006 with permission from Elsevier.

Ito et al subsequently confirmed that NAC prolongs the reduced lifespan of *Atm*$^{-/-}$ mice.[19] In this study, NAC treatment extended the animal survival from approximately 5 month to 10.7 month. Thus, both studies showed that NAC prolongs the longevity of *Atm*$^{-/-}$ mice by about 5-6 month.[14,19]

Several laboratories including ours looked into a possible mechanism of lymphoma prevention by NAC. An attractive explanation is that by inhibiting ROS NAC reduces oxidative DNA damage and, in turn, DNA strand break formation and deleterious genome rearrangements. In support of that, NAC was shown to reduce levels of intracellular ROS and chromosomal aberrations in cultured MEFs from *Atm*$^{-/-}$ mice [19] and oxidative DNA damage and resulting DNA deletions in tissues from *Atm*$^{-/-}$ mice.[33] In addition, NAC inhibits aberrant V(D)J recombination and restores T-cell development resulting in normalized thymic cellularity and increased counts of CD4 single positive T-cells.[19] Thus, NAC appears to reduce lymphomagenesis by reducing inherent genomic instability. Because NAC has a long history of safety and efficacy in humans,[39,40] it is not unreasonable to think that NAC can find its application in disease prevention in AT patients.

EUK-189

EUK-189, a salen-manganese compound with catalase and superoxide dismutase activities, has been previously shown to be neuroprotective in animal models characterized by oxidative damage.[44,45] EUK-189 was delivered via an osmotic pump implanted subcutaneously. EUK-189 improved *Atm*$^{-/-}$ mouse performance on a rotating rod and showed a trend towards prolonged life span ($p = 0.08$).[38] When the study was terminated at 5 months, 56% of EUK-189-treated and only 31% of vehicle-treated *Atm*$^{-/-}$mice were still alive.[38]

Tempol

Tempol (4-hydroxy-2,2,6,6-tetramethylpiperidine-N-oxyl) is a stable nitroxide free radical and superoxide dismutase mimetic.[46–48] Tempol detoxifies oxygen metabolites, oxidizes redox-active trace metal ions, reduces quinone radicals and, in biological systems, is itself reduced by GSH and ascorbic acid.[49–52] Tempol mixed in a mouse chow was chronically administered to *Atm*$^{-/-}$ mice either from fertilization or from weaning.[18] Tempol had no effect, when the treatment was started from fertilization, but increased the life span by 2-fold, when treatment was begun from weaning. Tempol reduced ROS levels, protein oxidation and restored mitochondrial membrane potential in thymocytes from *Atm*$^{-/-}$ mice. In contrast to NAC, tempol reduced the cell number in the thymus.

CTMIO

Like tempol, CTMIO (5-carboxy-1,1,3,3-tetramethylisoindolin-2-yloxyl) belongs to a class of stable nitroxide free radicals.[46–48] The effect of CTMIO intake via drinking water was examined in *Atm*$^{-/-}$ mice.[25] The treatment was started immediately after weaning. CTMIO prolonged animal survival by about 3-fold, improved performance on a narrow beam and reduced oxidative damage in the cerebellum.

Conclusion

Generation of *Atm* deficient mice allowed researchers to study AT in different cells and tissues. Most importantly, it allowed testing new therapeutic approaches, which would be impossible in human AT patients. As a potential means to prevent AT, antioxidant intake was examined in *Atm* deficient mice. Antioxidants, such as NAC, EUK-189, tempol and CTMIO, had some beneficial effect in extending the reduced life span, reducing lymphomagenesis and correcting neurobehavioral deficits in *Atm* deficient mice. Of the tested antioxidants, only NAC has a well-established safety and efficacy in humans. Therefore, it has a strong potential to become an antioxidant of choice to reduce the risk of cancer and/or neurological defects in AT.

An ongoing clinical trial in pediatric AT patients, where a cocktail of antioxidants including NAC is employed (personal communication with Dr. G. Berry, Thomas Jefferson University

Medical College, also see http://www.treat-at.org), will provide some information as to whether or not antioxidant therapy can slow down disease progression in AT.

References

1. Gatti RA, Lange E, Rotman G et al. Genetic haplotyping of ataxia telangiectasia families localizes the major gene to an approximately 850 kb region on chromosome 11q23.1. Int J Radiat Biol 1994; 66:S57-62.
2. Savitsky K, Bar-Shira A, Gilad S et al. A single ataxia telangiectasia gene with a product similar to PI-3 kinase. Science 1995; 268:1749-1753.
3. Pecker I, Avraham KB, Gilbert DJ et al. Identification and chromosomal localization of Atm, the mouse homolog of the ataxia telangiectasia gene. Genomics 1996; 35:39-45.
4. Barlow C, Hirotsune S, Paylor R et al. Atm-deficient mice: a paradigm of ataxia telangiectasia. Cell 1996; 86:159-171.
5. Elson A, Wang Y, Daugherty CJ et al. Pleiotropic defects in ataxia telangiectasia protein-deficient mice. Proc Natl Acad Sci USA 1996; 93:13084-13089.
6. Xu Y, Ashley T, Brainerd EE et al. Targeted disruption of ATM leads to growth retardation, chromosomal fragmentation during meiosis, immune defects and thymic lymphoma. Genes Dev 1996; 10:2411-2422.
7. Borghesani PR, Alt FW, Bottaro A et al. Abnormal development of purkinje cells and lymphocytes in atm mutant mice. Proc Natl Acad Sci USA 2000; 97:3336-3341.
8. Spring K, Cross S, Li C et al. Atm knock-in mice harboring an in-frame deletion corresponding to the human ATM 7636del9 common mutation exhibit a variant phenotype. Cancer Res 2001; 61:4561-4568.
9. Xu Y. ATM in lymphoid development and tumorigenesis. Adv Immunol 1999; 72:179-189.
10. Eilam R, Peter Y, Elson A et al. Selective loss of dopaminergic nigro-striatal neurons in brains of atm-deficient mice. Proc Natl Acad Sci USA 1998; 95:12653-12656.
11. Spring K, Ahangari F, Scott SP et al. Mice heterozygous for mutation in Atm, the gene involved in ataxia telangiectasia, have heightened susceptibility to cancer. Nat Genet 2002; 32:185-190.
12. Hecht F, Hecht BK. Cancer in ataxia telangiectasia patients. Cancer Genet Cytogenet 1990; 46:9-19.
13. Taylor AM, Metcalfe JA, Thick J et al. Leukemia and lymphoma in ataxia telangiectasia. Blood 1996; 87:423-438.
14. Reliene R, Schiestl RH. Antioxidant N-acetyl cysteine reduces incidence and multiplicity of lymphoma in Atm deficient mice. DNA Repair (Amst) 2006; 5:852-859.
15. Liyanage M, Weaver Z, Barlow C et al. Abnormal rearrangement within the alpha/delta T-cell receptor locus in lymphomas from Atm-deficient mice. Blood 2000; 96:1940-1946.
16. Reliene R, Schiestl RH. Differences in animal housing facilities and diet may affect study outcomes-a plea for inclusion of such information in publications. DNA Repair (Amst) 2006; 5:651-653.
17. Westphal CH, Rowan S, Schmaltz C et al. atm and p53 cooperate in apoptosis and suppression of tumorigenesis, but not in resistance to acute radiation toxicity. Nat Genet 1997; 16:397-401.
18. Schubert R, Erker L, Barlow C et al. Cancer chemoprevention by the antioxidant tempol in atm-deficient mice. Hum Mol Genet 2004; 13:1793-1802.
19. Ito K, Takubo K, Arai F et al. Regulation of reactive oxygen species by atm is essential for proper response to DNA double-strand breaks in lymphocytes. J Immunol 2007; 178:103-110.
20. Petiniot LK, Weaver Z, Barlow C et al. Recombinase-activating gene (RAG) 2-mediated V(D)J recombination is not essential for tumorigenesis in Atm-deficient mice. Proc Natl Acad Sci USA 2000; 97:6664-6669.
21. Chao C, Yang EM, Xu Y. Rescue of defective T-cell development and function in atm$^{-/-}$ mice by a functional TCR alpha beta transgene. J Immunol 2000; 164:345-349.
22. Lumsden JM, McCarty T, Petiniot LK et al. Immunoglobulin class switch recombination is impaired in atm-deficient mice. J Exp Med 2004; 200:1111-1121.
23. Petiniot LK, Weaver Z, Vacchio M et al. RAG-mediated V(D)J recombination is not essential for tumorigenesis in atm-deficient mice. Mol Cell Biol 2002; 22:3174-3177.
24. Kuljis RO, Xu Y, Aguila MC et al. Degeneration of neurons, synapses and neuropil and glial activation in a murine atm knockout model of ataxia telangiectasia. Proc Natl Acad Sci USA 1997; 94:12688-12693.
25. Gueven N, Luff J, Peng C et al. Dramatic extension of tumor latency and correction of neurobehavioral phenotype in atm-mutant mice with a nitroxide antioxidant. Free Radic Biol Med 2006; 41:992-1000.
26. Reliene R, Fleming SM, Chesselet MF et al. Effects of antioxidants on cancer prevention and neuromotor performance in Atm deficient mice. Food Chem Toxicol 2008; 46:1371-1377.

27. Eilam R, Peter Y, Groner Y et al. Late degeneration of nigro-striatal neurons in ATM$^{-/-}$ mice. Neuroscience 2003; 121:83-98.
28. Chen P, Peng C, Luff J et al. Oxidative stress is responsible for deficient survival and dendritogenesis in purkinje neurons from ataxia telangiectasia mutated mutant mice. J Neurosci 2003; 23:11453-11460.
29. Hande MP, Balajee AS, Tchirkov A et al. Extra chromosomal telomeric DNA in cells from Atm(−/−) mice and patients with ataxia telangiectasia. Hum Mol Genet 2001; 10:519-528.
30. Bishop AJ, Barlow C, Wynshaw-Boris AJ et al. Atm deficiency causes an increased frequency of intrachromosomal homologous recombination in mice. Cancer Res 2000; 60:395-399.
31. Reichenbach J, Schubert R, Schindler D et al. Elevated oxidative stress in patients with ataxia telangiectasia. Antioxid Redox Signal 2002; 4:465-469.
32. Barlow C, Dennery PA, Shigenaga MK et al. Loss of the ataxia telangiectasia gene product causes oxidative damage in target organs. Proc Natl Acad Sci USA 1999; 96:9915-9919.
33. Reliene R, Fischer E, Schiestl RH. Effect of N-acetyl cysteine on oxidative DNA damage and the frequency of DNA deletions in atm-deficient mice. Cancer Res 2004; 64:5148-5153.
34. Quick KL, Dugan LL. Superoxide stress identifies neurons at risk in a model of ataxia telangiectasia. Ann Neurol 2001; 49:627-635.
35. Kamsler A, Daily D, Hochman A et al. Increased oxidative stress in ataxia telangiectasia evidenced by alterations in redox state of brains from atm-deficient mice. Cancer Res 2001; 61:1849-1854.
36. Stern N, Hochman A, Zemach N et al. Accumulation of DNA damage and reduced levels of nicotine adenine dinucleotide in the brains of atm-deficient mice. J Biol Chem 2002; 277:602-608.
37. Barzilai A, Rotman G, Shiloh Y. ATM deficiency and oxidative stress: a new dimension of defective response to DNA damage. DNA Repair (Amst) 2002; 1:3-25.
38. Browne SE, Roberts LJ, 2nd, Dennery PA et al. Treatment with a catalytic antioxidant corrects the neurobehavioral defect in ataxia telangiectasia mice. Free Radic Biol Med 2004; 36:938-942.
39. Miller LF, Rumack BH. Clinical safety of high oral doses of acetylcysteine. Semin Oncol 1983; 10:76-85.
40. Kelly GS. Clinical applications of N-acetylcysteine. Altern Med Rev 1998; 3:114-127.
41. Webb WR. Clinical evaluaton of a new mucolytic agent, acetyl-cysteine. J Thorac Cardiovasc Surg 1962; 44:330-343.
42. Prescott LF, Park J, Ballantyne A et al. Treatment of paracetamol (acetaminophen) poisoning with N-acetylcysteine. Lancet 1977; 2:432-434.
43. De Flora S, Izzotti A, D'Agostini F et al. Mechanisms of N-acetylcysteine in the prevention of DNA damage and cancer, with special reference to smoking-related end-points. Carcinogenesis 2001; 22:999-1013.
44. Melov S, Doctrow SR, Schneider JA et al. Lifespan extension and rescue of spongiform encephalopathy in superoxide dismutase 2 nullizygous mice treated with superoxide dismutase-catalase mimetics. J Neurosci 2001; 21:8348-8353.
45. Doctrow SR, Huffman K, Marcus CB et al. Salen-manganese complexes as catalytic scavengers of hydrogen peroxide and cytoprotective agents: structure-activity relationship studies. J Med Chem 2002; 45:4549-4558.
46. Mitchell JB, Samuni A, Krishna MC et al. Biologically active metal-independent superoxide dismutase mimics. Biochemistry 1990; 29:2802-2807.
47. Hahn SM, Krishna CM, Samuni A et al. Potential use of nitroxides in radiation oncology. Cancer Res 1994; 54:2006s-2010s.
48. Damiani E, Kalinska B, Canapa A et al. The effects of nitroxide radicals on oxidative DNA damage. Free Radic Biol Med 2000; 28:1257-1265.
49. Branca M, Denurra T, Turrini F. Reduction of nitroxide free radical by normal and G6PD deficient red blood cells. Free Radic Biol Med 1988; 5:7-11.
50. Krishna MC, Grahame DA, Samuni A et al. Oxoammonium cation intermediate in the nitroxide-catalyzed dismutation of superoxide. Proc Natl Acad Sci USA 1992; 89:5537-5541.
51. Krishna MC, Russo A, Mitchell JB et al. Do nitroxide antioxidants act as scavengers of O_2-. or as SOD mimics? J Biol Chem 1996; 271:26026-26031.
52. May JM, Qu ZC, Juliao S et al. Ascorbic acid decreases oxidant stress in endothelial cells caused by the nitroxide tempol. Free Radic Res 2005; 39:195-202.
53. Reliene R, Schiestl RH. Antioxidants suppress lymphoma and increase longevity in Atm-deficient mice. J Nutr 2007; 137:229S-232S.

INDEX

I

L

M

N

O

P

R

S

T

V

For Product Safety Concerns and Information please contact our EU representative GPSR@taylorandfrancis.com Taylor & Francis Verlag GmbH, Kaufingerstraße 24, 80331 München, Germany

Printed and bound by CPI Group (UK) Ltd, Croydon, CR0 4YY
14/09/2026
02216223-0014